Nonsurgical Lip and Eye Rejuvenation Techniques

Gabriella Fabbrocini
Maria Pia De Padova • Antonella Tosti
Editors

Nonsurgical Lip and Eye Rejuvenation Techniques

Springer

Editors
Gabriella Fabbrocini
Federico II
University of Naples
Naples
Italy

Antonella Tosti
University of Miami
Miller School
Miami, FL
USA

Maria Pia De Padova
Department of Dermatology
Nigrisoli Hospital
Bologna
Italy

ISBN 978-3-319-23269-0 ISBN 978-3-319-23270-6 (eBook)
DOI 10.1007/978-3-319-23270-6

Library of Congress Control Number: 2016944467

Printed on acid-free paper

This Springer imprint is published by Springer Nature
The registered company is Springer International Publishing AG Switzerland

Preface

The average age of the population is constantly rising all over the world, in particular in the industrialized nations. Therefore, the geriatric population represents the fastest growing segment of Western countries. The effects of human aging are primarily visible in the skin, skin laxity, and changes in skin pigmentation. Since humans have always been fascinated by conserving youth, there has been an extraordinary spreading of both surgical and nonsurgical cosmetic procedures in the last two decades. Particularly, according to statistics from the American Society for Aesthetic Plastic Surgery, since 1997, there has been an increase of 444 % in the total number of cosmetic procedures in the United States with surgical and nonsurgical ones increased by 119 % and 726 %, respectively. Therefore, understanding the mechanisms of skin aging is the key point in order to correctly and effectively reduce the signs of aging on the skin through the use of proper and safe intervention modalities. In this context, the face, particularly in the perioral and the periorbital areas are key areas of intervention.

This text was based on authors' experience and careful review of the literature.

The perioral and the periorbital regions are complex and dynamic parts of the face and it is necessary to know their peculiar anatomic components for the correct choice of the procedures. Successful rejuvenation often requires a combination of minimally invasive modalities to fill dents and hollows, resurface rhytides, improve pigmentation, and smooth the mimetic muscles of the face without masking facial expression.

Possible procedures include botulinum toxin, facial filler, skin needling, chemical peelings, radiofrequency, biorivitalization, ablative and nonablative laser, PRP, and suture suspension technique that can be variably combined to provide enhanced outcomes.

Many aesthetic procedures for lip wrinkles are available: *static wrinkles* can be treated through facial skin resurfacing, laser, mechanical dermabrasion, skin needling, chemical peels, and soft tissue fillers; for *dynamic wrinkles*, BOTOX® injections can be very useful. This wide selection of techniques allows us to choose those with higher efficacy, minimal adverse effects, and short downtime.

This book teaches beginners and nonbeginners how to chose and perform at the best each procedure with great attention to prevention and management of possible side effects.

Naples, Italy — Gabriella Fabbrocini
Bologna, Italy — Maria Pia De Padova
Miami, FL, USA — Antonella Tosti
June 2016

Contents

Introduction: Anatomy of the Lips and Eye

1

Corinna Rigoni

1.1 Anatomy of the Eyelids

The eyelids are highly specialized structures with peculiar anatomic components. The ocular globes are allocated in two symmetrically bony cavities called orbits, consisting of seven bones that develop the orbital walls. The roof is composed mostly of the orbital plate of the frontal bone and posteriorly of a minor part of the sphenoid bone. The lateral wall comprises the orbital surface of the zygomatic bone and the sphenoid bone. The floor is composed of the orbital plate of the maxilla anterolaterally of the zygomatic bone and posteriorly of the palatine bone. The medial wall consists of the ethmoid, frontal, lacrimal, and sphenoid bone. The eyelid skin, which is less than 1 mm thick, is a thin epidermis constructed from a stratified epithelium of 67 cell layers. The dermis contains elastic fibers, blood vessels, lymphatics, and nerves. The underlying fat is scant or not present in the subcutaneous tissue, where the hair follicles and pilosebaceous glands are located. The apocrine glands of Moll are located near the lid margin, and the sebaceous glands of Zeiss are associated with the follicles of the eyelashes. The eyelids function to protect the eye globe from local and external injuries. Furthermore, they regulate the light that reaches the eye and uniformly distribute the tear film, mucus, and oil during blinking, of great importance for the health of the cornea. The eyelids are divided into upper and lower eyelids, which are similar but with different characteristics mainly in the lid retractor arrangement. The space between the open lids is known as the palpebral fissure, which measures 7–12 mm, while the normal excursion of the lids is 14–17 mm. In the normal adult fissure, the highest point of the upper lid is just nasal to the center of the pupil, while the lowest point of the lower lid is just temporal to the center of the pupil. In youths, the upper lid margin rests at the upper limbus, whereas in adults it rests 1.5 mm below the limbus. The lower eyelid margin rests at the level of the lower limbus. The lateral canthal angle is 2 mm higher than the medial canthal angle in Europeans, but is 3 mm higher in Asians. The distance from the medial canthus to the midline of the nose is approximately 15 mm. The lateral canthus lies directly over the sclera, and the medial canthus is separated from the eye by the lacrimal lake and caruncle, a yellowish tissue containing sebaceous and sweat glands. The lid margins are 2 mm wide and form the junction between the skin and the conjunctiva, the mucous membrane of the lids. They meet at the gray line, near the posterior edge of the lid margin, the junction of the anterior and posterior lamellae of the lids. The eyelashes are located anteriorly and the openings of the meibomian glands posteriorly. There are approximately 100–150 eyelashes on the upper lid and about 50–75 on the lower. The follicular structure of eyelashes includes the

C. Rigoni, MD
Milan, Italy
e-mail: c.rigoni@twtnet.com

G. Fabbrocini et al. (eds.), *Nonsurgical Lip and Eye Rejuvenation Techniques*,
DOI 10.1007/978-3-319-23270-6_1

sebaceous (Zeiss) and sweat (Moll) glands, while the tarsal glands (Meibomian) open posteriorly to the lid margin. The tears that appear at the tips of the small papillae are drained from the surface of the eyes through the openings by a pump mechanism. The lacrimal secretory system controls the amount of tears and is divided into the basic and reflex secretors. The basic secretor is composed of three sets of glands. (1) Conjunctival, tarsal, and limbal mucin-secreting goblet cells; the overlying aqueous layer is spread more uniformly because of this inner layer (precorneal tear film). (2) The accessory lacrimal exocrine glands of Krause and Wolfring, located in the subconjunctival tissue. (3) The oil-producing Meibomian glands and the palpebral glands of Zeiss and Moll. The reflex secretor is divided into two parts by the lateral horn of the levator palpebrae superioris.

The first fold of the upper eyelid is represented by the superior palpebral sulcus, 9–10 mm (individual and racial variations) above the lid margin, and represents the junction of the levator palpebrae superioris with the orbital septum and the fibrous insertion of the levator aponeurosis into the skin.

There is a thin fascial layer between the skin and the orbicularis oculi muscle, with no fat tissue. The eyelid normally is located at the superior border of the tarsus, and the skin below the lid is attached to the underlying tarsus with the levator aponeurosis, which has projections anteriorly through the pretarsal orbicularis to the skin and posteriorly to the inferior portion of the anterior tarsus. The skin of the upper eyelid is more freely movable because of the lack of superior aponeurotic attachments and underlying orbital septum.

The second layer of the eyelid is the orbicularis oculi muscle, which is divided into orbital and palpebral parts that function independently. The orbital part is a voluntary muscle while the palpebral part is both voluntary and involuntary. The orbital portion extends in a wide, circular fashion around the orbit, interdigitating with other muscles of facial expression. It has a curved origin from the medial orbital margin, being attached to the superomedial orbital margin, maxillary process of the frontal bone, medial palpebral ligament, frontal process of the maxilla, and inferomedial orbital margin. The palpebral portion is further subdivided into pretarsal and preseptal portions. The preseptal orbicularis muscle covers the orbital septum and originates medially from a superficial and deep head associated with the medial palpebral ligament. The fibers from the upper and lower eyelid join laterally to form the lateral palpebral raphe, which is attached to the overlying skin. The pretarsal portion lies anterior to the tarsus, with a superficial and deep head of origin intimately associated with the medial palpebral ligament. Fibers run horizontally and laterally to extend deep to the lateral palpebral raphe, to insert in the lateral orbital tubercle through the intermediary of the lateral canthal tendon. The peripheral fibers sweep across the eyelid over the orbital margin in a series of concentric loops, the more central ones forming almost complete rings, interdigitating with other muscles of facial expression. In the upper lid the orbital part extends as far as the forehead, covers the corrugator supercilii muscle, and continues laterally over the anterior temporal fascia.

The third layer of the lids in the upper portion is the orbital septum, a fascial membrane that separates the eyelid structures from the deeper orbital structures, and attaches to the orbital margin a thickening called the arcus marginalis, the point of confluence for the facial bone periosteum and the periorbita. With age, the septum weakens and bulging of the orbital fat pad becomes visible. Its removal is important in blepharoplastic surgery.

The fourth layer of the upper lid is the postseptal fat pad, contained within the orbit by the orbital septum.

In the lower lid, the orbital part lies on the origins of the elevator muscles of the upper lip and nasal ala, and continues to cover partially the masseter muscle (Figs. 1.1 and 1.2).

Asians have different periorbital anatomic characteristics, the eyelid being one of the most prominent features of the face. Moreover, there is also a wide variety of eyelid shapes, mostly with regard to the presence and location of the supratarsal fold and/or presence of an epicanthal fold. The most obvious feature of Oriental eyelids is the absent or very low supratarsal fold with

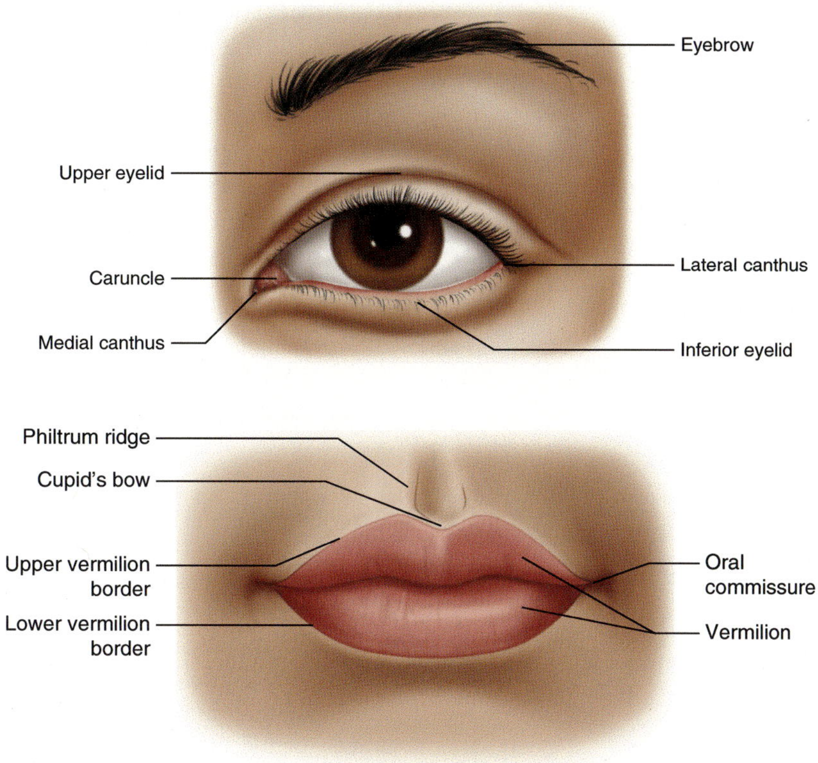

Fig. 1.1 Eyelids: (*1*) Eyebrow, (*2*) Upper eyelid, (*3*) Medical cantus, (*4*) Caruncle, (*5*) Lateral cantus, (*6*) Inferior eyelid. Lips: (*1*) Philtrum ridge, (*2*) Cupid's bow, (*3*) Upper vermilion border, (*4*) Vermilion, (*5*) Oral commessure, (*6*) Lower vermilion border

relatively "full" periorbital tissues. Only a very small percentage of Orientals have a manifest supratarsal fold. In fact, at this location in other ethnic groups the levator aponeurosis sends fibers to the overlying skin, anchoring it down to the eyelid, creating the fold. In Orientals this fusion is scarce, making the supratarsal fold closer to the eyelid edge. Because the septal-levator fusion is so low on the eyelid, retroseptal fat descends in the fold and creates an impression of a fuller upper eyelid. A submuscularis fibroadipose tissue layer and a more lowly positioned transverse ligament were recently identified and found exclusively in the Asian eye (Fig. 1.3).

1.1.1 Conjunctiva

The conjunctiva is a smooth, translucent mucous membrane of stratified columnar epithelium, situated on the inferior surface of the tarsus, from the mucocutaneous junction of the lid margin to the tarsal plate border. The conjunctiva is reflected at the fornix on the globe as the bulbar conjunctiva. Tarsal conjunctiva is adherent to the tarsus, while a submucosal lamina propria underlies orbital palpebral conjunctiva and allows dissection from the vascular Müller muscle. The ciliary muscles of Riolano are situated anterior to the tarsus and near the cilia.

1.1.2 Blood Supply

The arteries of the eyelids develop from the internal carotid artery through the ophthalmic artery and the external carotid artery through the facial and superficial temporal branches. The branches of the internal carotid are laterally, the lacrimal artery and medially, the supratrochlear and medial palpebral branches of the ophthalmic artery.

The veins of the eyelids are called the pretarsal and posttarsal veins. The pretarsals are superficial and are connected medially to the angular vein and lateroposteriorly to the superficial temporal and lacrimal veins. The posttarsals are deeper, and connect the orbital veins with the deeper branches of the anterior facial vein and the pterygoid plexus.

The lymphatics of the eyelids, like the veins, have pretarsal and posttarsal systems. The lateral vessels drain the lateral areas of the lids and the deeper vessels the conjunctiva of the upper folds and lacrimal glands, which drain into the superficial and deep preauricular nodes.

1.1.3 Nerves

The temporal branch of the facial nerve innervates the upper region while the zygomatic branch of the facial nerve innervates the lower region. Sensory innervation of the eyelids is supplied by terminal branches of the ophthalmic and maxillary divisions of the trigeminal nerve. Within the superior orbit, the frontal branch of the ophthalmic division of the trigeminal nerve arrives anteriorly between the periorbita of the roof and the levator muscle. Here, it divides into

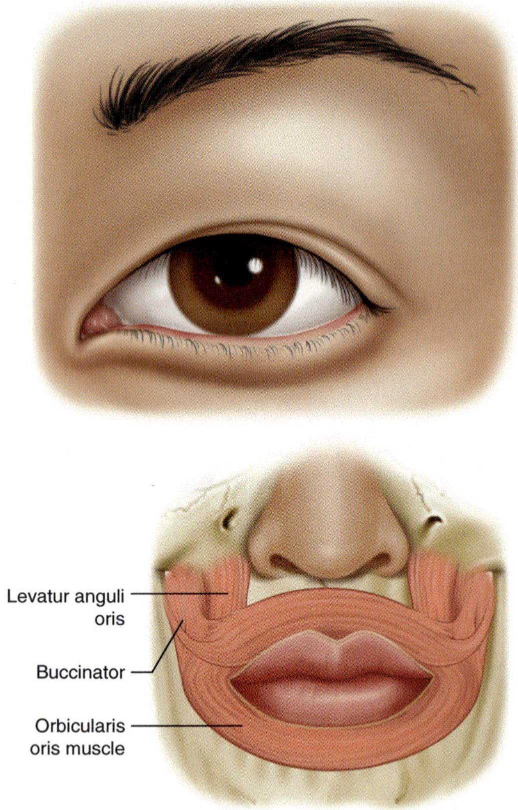

Fig. 1.2 Eyelids: Oriental upper eyelid. Lips: (*1*) Orbicularis oris muscle, (*2*) Levatur anguli oris, (*3*) Buccinator

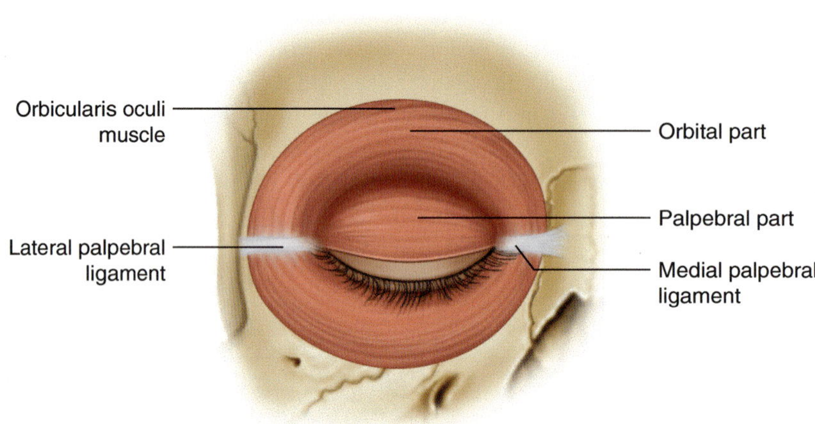

Fig. 1.3 (*1*) Orbicularis oculi muscle: (*a*) orbital part; (*b*) palpebral part, (*2*) Lateral palpebral ligament, (*3*) Medial palpebral ligament eyelids

a larger supraorbital nerve and a smaller supratrochlear nerve. Terminal branches of these nerves supply sensation to the upper eyelid and forehead.

1.2 Anatomy of the Lips

The lips are subjected to numerous movements, so their aspect varies according to movement. Furthermore, from their shape (motion) we can guess whether a person is happy or sad. Their function, together with the mouth and the oral cavity, is supported by a complex system of structures and muscles to participate in the process of mastication, speaking, and especially nonverbal communication.

They are so pliable and elastic that they are capable of wide excursions of movement.

The lips form the mouth and surround the oral cavity. They lie in the central portion of the inferior third of the face. The upper lip corresponds superiorly to the inferior margin of the base of the nose and extends laterally to the nasolabial fold, and inferiorly to the free edge of the vermilion border. The lower lip extends from the superior free vermilion edge superiorly, to the commissures laterally, and the mandible inferiorly. The upper and lower lips join at the oral commissures. Inferiorly the limit of the lips in the central region is the mentolabial sulcus, which intraorally corresponds to the depth of the gingivolabial sulcus.

From the anatomic viewpoint, the philtrum and its pillars belong to the upper lip. The philtrum lies in the central region and extends from the base of the nose to the upper lip border. It is a depression between two raised vertical columns of tissue known as the pillars. The surface of the lips is composed of hairy skin, vermilion border, vermilion, and oral mucosa. The shape of the lips varies with age and ethnicity. The vermilion is the red part of the lips and is covered with a modified mucous membrane, which continues with the oral mucosa of the gingivolabial sulcus, and is dry as it contains no salivary, sweat, or oil (sebaceous) glands. The vermilion border is the edge of clearer skin that borders the vermilion. The Cupid's bow is considered the contour of the line formed by the vermilion border in the central region of the upper lip. The philtrum is formed by a combination of longitudinal collagen condensations supported by a rich elastic tissue component and interdigitating orbicularis oris muscle fibers.

The oral mucosa consists of stratified squamous non-keratinized epithelium and covers the part inside the oral cavity of the lips. The oral commissure represents the point at which the lateral borders of the vermilion of the upper and lower lips join.

The external surface of the lips is rich in sebaceous glands, whose secretion prevents dryness and desquamation. The labial glands are minor salivary glands situated between the mucous membrane and the orbicularis oris muscles around the orifice of the mouth. The labial glands are circular in form and about the size of small peas; their ducts open by minute orifices on the mucous membrane

The perioral orbicularis oris musculature, the intrinsic and circumferential muscle of the lip, provides the center of the coordination of muscular activity. The orbicularis oris muscle, a voluntary, mimic striated muscle, has no bony attachment and is not supported by bone or cartilage, and has a sphinteric function. Into this muscle insert the antagonistic and synergistic elevator, depressor, and retractor muscle groups that create coordination between contraction and relaxation of the movements of the buccinator, levator anguli oris, depressor anguli oris, zygomaticus major, and risorius that insert into the modiolus. This is formed by several retractor muscles converging to act on the angle of the mouth. Lip elevator muscles insert into the upper lip: levator labii superioris, levator anguli oris, levator labii superioris alaeque nasi, and zygomaticus minor and major. The lip depressors are the: depressor labii inferioris, mentalis, and platysma.

The motor innervation to the perioral musculature uniformly comes from the seventh cranial nerve, the facial nerve. The buccal and marginal branches primarily supply innervation to the perioral musculature. The fibers supply the majority of the muscles of the face from their undersurface. The lips are abundantly provided with sensitive nerve endings. Sensory innervation is supplied to the upper lip by the infraorbital branch of the maxillary division of the trigeminal nerve. The nerve runs beneath the levator labii superioris and superficial to the levator anguli oris to supply the lateral nasal sidewall, ala, columella, medial cheek, and upper lip.

The mental nerve, a branch of the mandibular nerve, innervates the lower lip.

The arterial vascularization of the lips, originating in the external carotid system, is supplied by the superior and inferior labial arteries that arise from each facial artery lateral to the commissure. Venous drainage occurs via the anterior facial vein and partly via the submental veins.

The lymphatic drainage of the lips occurs through submandibular and submental nodes. Lymphatic drainage from the upper lip is unilateral except for the midline, where some drainage is available to the submental nodes. The upper lip and lateral lower lip drain to the submandibular nodes.

Bibliography

Anatomy of the Eyelids

Chau-Jin W (2009) Aesthetic surgery in Asians. Semin Plast Surg 23:5–15
Dailey RA, Wobig JL (1992) Eyelid anatomy. J Dermatol Surg Oncol 18:1023–1027
Kiranantawat K, Suhk JH, Nguyen AH (2015) The Asian eyelid: relevant anatomy. Semin Plast Surg 29: 158–164
Salasche SJ et al (1988a) Surgical anatomy of the skin. Appleton &Lange, Norwalk, pp 223–240
Zide BM, Jelks GW (1985) Surgical anatomy of the orbit. Raven, New York, pp 21–39

Anatomy of the Lips

Carey JC et al (2009) Elements of morphology: standard terminology for the lips, mouth, and oral region. Am J Med Genet part A. 149A(1):77–92
Salasche SJ et al (1988b) Surgical anatomy of the skin. Appleton &Lange, Norwalk, pp 223–240
Zugerman C (1986) The lips: anatomy and differential diagnosis. Cutis 38:116–120

Rejuvenative Outcomes for the Lip and Eye Area

2

Giuseppe Monfrecola and Matteo Megna

Abbreviations

ECM	Extracellular matrix
GAGs	Glycosaminoglycans
IL	Interleukin
MMPs	Matrix metalloproteinases
ROS	Reactive oxygen species
TNF	Tumor necrosis factor
UV	Ultraviolet rays
VIS	Visible

2.1 Introduction

The average age of the population is constantly rising all over the world, in particular in the industrialized nations [7, 8]. Therefore, the geriatric population represents the fastest growing segment of Western countries. It has been estimated that the elderly will constitute up to 25 % of the US population by 2025 and up to 34 % of the European population by 2050 [31, 62] and that the average life span is expected to extend another 10 years by 2050 worldwide [7]. The effects of human aging are also primarily visible in the skin with alterations such as atrophy, increased wrinkling, sagging, skin laxity, and changes in skin pigmentation [26, 40].

In this context, it is well known that humans have always been fascinated by conserving youth. Indeed, there has been an extraordinary spreading of both surgical and nonsurgical aesthetic and cosmetic procedures in the last two decades [28]. Particularly, according to statistics from the American Society for Aesthetic Plastic Surgery, since 1997, there has been an increase of 444 % in the total number of cosmetic procedures in the United States with surgical and nonsurgical ones being increased by 119 % and 726 %, respectively [1].

Therefore, understanding the mechanisms of skin aging is the key point in order to correctly and effectively counteract and reduce the time effects on the skin through the use of proper and safe intervention modalities.

2.2 Lip and Eye Area Aging

We can perceive the age of people from the appearance of their face as this is the part of the body which is the most exposed to environmental factors such as ultraviolet (UV) radiation above all; therefore it is not surprising that it represents one of the body areas where the signs of skin aging initially appear. For example, wrinkles, which constitute the most common and worrying sign of skin aging, are usually particularly concentrated around the eyes and lip [66].

G. Monfrecola, MD (✉) • M. Megna, MD
Section of Dermatology, Department of Clinical Medicine and Surgery, University of Naples Federico II, Via Pansini 5, Naples 80131, Italy
e-mail: monfreco@unina.it; mat24@libero.it

G. Fabbrocini et al. (eds.), *Nonsurgical Lip and Eye Rejuvenation Techniques*,
DOI 10.1007/978-3-319-23270-6_2

As regards the eye area, it is well known that few of the first signs of aging appearing in the late 20s and early 30s are usually located around the eyes. In this area, the skin undergoes numerous morphological and structural changes leading to the typical aging alterations observed in the orbital region such as brow ptosis, dermatochalasis, blepharochalasis, periorbital wrinkles, fat pad, malar bags, etc. [29, 49]. The considerable enhancement in skin thinning is involved in the appearance of dynamic rhytids at the lateral canthi known as crow's feet, whereas increased laxity on the upper lid leads to hooding and occasionally pseudoptosis, a condition generally known as dermatochalasis [3, 49]. Moreover, lower lid skin and orbital septum laxity are able to guide to the formation of bags which may also be favored by edema and skin stretching (malar bags). When both the skin and the orbicularis muscle are involved, the presence of redundant folds of loose skin, muscle, fat, and interstitial edema which extend from beyond the lateral cheek often past the midpupillary line, or even from canthus to canthus, may develop defining a condition commonly known as festoons [22, 23]. Apart from skin alterations, also muscle and subcutaneous tissue modifications contribute to the development of other noticeable signs of aging in the periorbital area. For example, contraction of the orbicularis muscle drives to changes in the overlying skin supporting the formation of the condition known as crow's feet; changes in fat amount and position are also strictly linked to aging variations observed in eye surrounding area, whereas important transformations in the laxity of the connective tissue structures and the canthal tendons may lead to a smaller appearance of the eyes, scleral show, or even ectropion [11, 17].

Apart from the periorbital area, lips, which are part of the aesthetic unit that involves the mouth and the perioral tissue, represent another face site particularly susceptible to manifest aging signs. While during puberty the lips become fuller because of the hypertrophy of the orbicularis muscle and glandular components, they progressively lose definition as a person ages, tending to become flatter and presenting also upper lip elongation. The oral commissures tend to descend and vertical wrinkles develop at or above the vermillion border due to skin thinning and orbicularis muscle atrophy [56]. Aging of this area is also characterized by perioral fine lines, marionette lines, and flattening of the cupid bow [71]. The dynamics of lip movement change with age too. The smile, for instance, gets narrower vertically and wider transversely [15]. Moreover, the consequences of the aging process are also the most evident along the mandible area where loss of subcutaneous fat tends to create a prejowl sulcus between the chin and sagging lower cheek and anterior to the masseter muscle [56].

2.2.1 Skin Aging Mechanisms

As already mentioned, aging affects the human face by provoking an array of microscopic and macroscopic complex volumetric changes [6]. These changes are exacerbated and/or accelerated by bad habits (e.g., smoking) and environmental factors. Therefore, both intrinsic and extrinsic factors are responsible for skin aging [18, 46], together leading to reduced structural integrity and loss of physiological function [46].

2.2.2 Intrinsic Aging

Intrinsic aging is defined as the amount of corporal changes that develop during the normal aging process affecting all body areas as a result of genetic factors [18]. As regards the skin, the intrinsic aging process leads to epidermal and dermal thinning [75]; intrinsically aged skin appears to be thin, dry, and transparent, presenting with fine wrinkles and irregular hair growth and touring out to be unable to sweat sufficiently [64]. As a consequence, skin effectiveness to act as a first barrier against environmental and external factors gradually decreases. Other cutaneous intrinsic alterations are linked to a reduction in the number of nerve endings and in the production of sex hormones which are responsible for decreased skin sensibility [67, 74]. Concerning the histopathological modifications, general atrophy of the extracellular matrix (ECM) with decreased elastin and disintegration of elastic fibers represents the most common features of intrinsically aged skin [45]. All these events may vary in relation to body site, differing also per ethnic group [14]. Moreover, in intrinsically aged skin, there is a decrease in vessel size without a significant difference in the vascular density [10]. However, even in subjects living strictly indoor all their life, skin that is aged only by intrinsic factors does not exist. Obviously, aged skin always reflects a variable impact of extrinsic aging, superimposed on the level of intrinsic aging [41].

2.2.2.1 Mechanisms of Intrinsic Aging

One of the major determinants of intrinsic aging is represented by reactive oxygen species (ROS), which are continuously produced inside our body as a result of the aerobic metabolism in the mitochondria [18]. Indeed, in the skin, about 1.5–5 % of the consumed oxygen is converted into ROS by intrinsic processes, with keratinocytes and fibroblasts being the main cutaneous producers [58]. The reactive superoxide anion radical (•O2−) is the principal ROS type formed in mitochondria, being able to harm numerous different cellular functions leading to nuclear and mitochondrial DNA damage, telomere shortening, protein glycosylation, lipid and protein oxidation, collagen and elastin degradation, downregulation of collagen synthesis, increased expression of matrix metalloproteinases (MMPs), as well as neovascularization [12, 41]. Moreover, not only ROS production increases with age, but also human skin cell ability to repair DNA damage steadily decreases over the years, potentiating ROS effects [59]. Apart from ROS production, other main factors which play an important role in intrinsic skin aging are the reduction in replicative ability of cells (cellular senescence) and the enhancement of ECM degradation. The replicative capacity of human cells decreases with time, and in the skin, this is particularly true for keratinocytes, melanocytes, and fibroblasts. Thus, senescent cells not able to undergo cellular division are found in higher levels in aged skin [16]. This is due to telomere shortening: with each cellular division, a small fragment of the telomere is definitively lost at the chromosome ends, and after 25–30 cellular divisions, they become so critically short that DNA loss during subsequent cell divisions leads to decline of somatic cell function, cell cycle arrest, and senescence [24]. Finally, an additional factor involved in intrinsic skin aging is represented by the increased expression of enzymes which act degrading ECM of the dermis; for example, an increase in MMP expression together with the reduction of the MMP inhibitors has been shown in aged fibroblast [51]. Notably, all these extracellular alterations of ECM may be triggered by ROS production [47].

2.2.3 Extrinsic Aging

Extrinsic aging is caused by external environmental factors such as solar radiation [25, 70], cigarette smoking [4], pollutants, etc. Particularly, UV exposure is believed to be the primary factor involved in extrinsic skin aging, through a process known as photoaging. This is especially true for exposed body sites such as the face. Indeed, about 80 % of facial aging is due to photoaging [21]. The rate and the intensity of UV radiation effects on skin aging are related to several factors such as frequency, duration, and intensity of solar exposure as well as the different phototypes [54], being more prominent in fair skin individuals (skin types I and II) and less noticeable in subjects with skin type III or higher [14, 63]. Photoaging is a cumulative process which shows a wide range of effects on the skin. Photoaged skin is commonly characterized by the presence of wrinkles, pigmented spots and pigmentation disorders, verrucous papules, dryness, telangiectasias, loss of elasticity, laxity, and rough-textured appearance [18, 20]. Particularly, the formation of wrinkles and small brown pigmented sharply demarcated spots, known as lentigines, seems to be the most common hallmark of photoaging; for these reasons their development mechanisms are discussed in detail in the following subheading. However, photoaging damage predominantly occurs in the connective tissue, also referred to as ECM whose most important and abundant structures being collagen, elastin, and glycosaminoglycans (GAGs), all essential to maintain the strength, the elasticity, and the hydration of the skin [55]. Indeed, regarding histopathological changes, progressive disorientation of dermal collagen and elastic fibers bundles is a common feature of photoaged skin. A significant increase in space between fiber bundles, thinning of fibers, and increased disorganization of fiber proteins are also present [50, 69]. Finally, photoaged skin is characterized by a loss of mature collagen, and basophilic degeneration of connective tissue, evidenced by denatured elastin fibers and collagen fibers [52].

Typical for a photoaged skin is the deposition of abnormal elastin with histological examination revealing deranged and highly branched elastic fibers that form aggregates of elastotic material formed by a combination of UV- or ROS-induced degradation of elastin and biosynthesis of amorphous and dysfunctional elastin and fibrillin [53]. Moreover, an age-dependent decrease in the cutaneous vascularity of sun-exposed facial area together with a reduction in vessel size and vascular number compared to younger skin is also reported [9].

2.2.3.1 Mechanisms of Extrinsic Aging

UV radiation is the main actor of extrinsic aging, being able to damage various cellular structures both directly and indirectly, thereby accelerating the aging process. An important role is played by UVB which is mainly responsible for direct cell damage. Indeed, even if a great amount of UVB is absorbed in the stratum corneum, attenuated UVB radiation also reaches viable epidermal cells [35], determining biological damage. Particularly, the most dangerous and critical type of biological damage is represented by DNA alteration [35, 57]. Actually, upon UVB cellular absorption, various DNA mutations may be set up through the formation of bonds between adjacent pyrimidines which cause the development of cyclobutane–pyrimidine dimmers and pyrimidine–pyrimidone (6-4) photoproducts [30]. Thus, mutated DNA and RNA bases are able to affect cellular protein synthesis and accumulation of unrepaired mutations can cause cell cycle arrest and apoptosis. However, at the same time, mutations can impair the cell apoptotic ability, enhancing skin malignancies' development. All these events are deeply influenced by the skin cell type, the cumulative UV dose, and the UV wavelength type which impact the final outcome [41]. Moreover, UV radiation is also able to induce biological damage and accelerate aging through indirect pathways with endogenous or exogenous photosensitizers that absorb UVA and even visible (VIS) wavelengths from both the sun and artificial sources [2, 60]. As a result, ROS such as singlet oxygen or direct photochemical changes to biomolecules may be performed [60]. Therefore, even if UVA and VIS radiation are less absorbed by epidermal components and hence penetrate deeper into the dermis, they should not be ignored as potential source of photoaging. In addition, aging modification is also indirectly caused by UV through ROS-induced damage which is able to stimulate the synthesis of MMPs [5]. Thus, other important factors of photodamage are also increased MMP overexpression and activity, being responsible to the degradation of dermal connective tissue [19]. Particularly, MMP upregulation is able to occur after low UV exposure doses, less than one minimal erythema dose [42]. Therefore, even daily exposures to a low-dose solar UV radiation below sunburn are thought to be sufficient to induce MMP upregulation and their related photoaging consequences such as degradation of skin collagen and elastic fibers above all. Notably, MMP production is not only induced by ROS but also by inflammatory cells (macrophages and neutrophils) which infiltrate the skin after UV-induced inflammatory effects [65]. Apart from ECM degradation through MMPs, UV-induced ROS are also able to damage GAGs, important structures to give support, strength, and flexibility to the connective tissue and keep the tissue hydrated [43]. For example, the most known member of GAG family, hyaluronic acid, is strongly reduced in the dermis after chronic UVB exposure [13]. Furthermore, UV is also able to increase the expression of fibromodulin, a small leucine-rich repeat protein which interacts with type I and II collagen fibrils, thereby affecting ECM metabolism through the alteration of the balance between collagen synthesis and degradation, leading to collagen deficiency observed in photoaged skin [39]. Therefore, as described above, UV radiation plays a major role in extrinsic aging (photoaging). Particularly, wrinkles and lentigine formation represent the two most classical examples of the key role played by UV radiation in determining the skin effects of extrinsic aging (photoaging). Indeed, regarding wrinkles, UV-induced degradation of skin collagen and elastic fibers through MMP activity is one of the main mechanisms involved in their formation [38]. Particularly, UVR-induced ROS are able to activate signaling kinases (activator protein-1 and MAPK signaling) which control the transcription of MMPs in epidermal keratinocytes and dermal fibroblasts [19]. Moreover, keratinocytes exposed to UVB radiation produce and secrete cytokines such as interleukin (IL)-1α, IL-6, and tumor necrosis factor (TNF-α), which stimulate epidermal keratinocytes and dermal fibroblasts and enhance MMP-1, MMP-2, MMP-9, and MMP-12 levels leading to wrinkle formation through damage of dermal collagen and elastic fibers [19, 20, 38, 42, 44]. Furthermore, UVB radiation is also able to induce MMP-1, MMP-3, and MMP-9 in normal human epidermis, whereas UVA stimulates the expression of MMP-1, MMP-2, and MMP-3 in fibroblasts [34]. In addition, UVB radiation may also contribute to wrinkle formation by inducing fibroblast elastase via cytokines released by exposed keratinocytes [72].

Consequently, several mechanisms are considered to be involved in wrinkle development such as the decrease of collagen and elastin fibers in dermis, the degradation of basement membrane at the dermal–epidermal junction, and a decrease in the three-dimensional organization of the ECM [20, 68]. Hence, UV radiation has been implicated in wrinkle formation through its exacerbation of the decline in tensile strength and elasticity and its ability to cause the degradation of the supporting structural components of the dermal ECM.

The key role of UV radiation in extrinsic aging is also showed by their strong involvement in the development of solar lentigines, another classical marker of skin aging together with wrinkles. These brown pigmented lesions may be induced by mutations of keratinocytes and melanocytes which both play a role in pigment formation and transfer. In this context, UV radiation is considered the principal actor in their formation process. Particularly, through its ability to induce mutations in cutaneous cells, UVB radiation is thought to be the extrinsic factor most responsible for pigment spot development [36, 37].

2.2.4 Aging: Intrinsic and Extrinsic Mechanism Overlap

Mechanisms of intrinsic aging and extrinsic aging share substantial overlap, both featuring DNA damage [27]. For example, critical shortening of telomeres, which cause cellular senescence and organism aging overall, is related to finite number of cell divisions, depending to passage of time in proliferative tissues which also characteristically increase after injury, including UV irradiation [32]. ROS production represents another common executor of both intrinsic and extrinsic aging, being strictly linked to DNA damage and senescence. Indeed, it is well known that ROS can be produced by both intrinsic aerobic metabolism [18] and extrinsic UV exposure [33]. Moreover, when a cell enters senescence, p53 functions such as enhanced DNA repair capacity and stimulation of antioxidant defenses cease [48], leaving viable but nonproliferative cells (e.g., dermal fibroblasts) in a state of chronic oxidative stress that promotes the pro-inflammatory environment characteristic of old skin, making them also more susceptible to UV radiation-induced damage [73]. As a result of these common alterations, it is not surprising that both intrinsic and extrinsic aging are able to determine some similar qualitative and quantitative changes in ECM, leading to loss of tensile strength and recoil capacity, wrinkle formation, dryness, impaired wound healing, and increased fragility [58]. Nevertheless, not all aging-related ECM modifications are analogous between intrinsic and extrinsic pathways; e.g., globally, intrinsically aged skin preferentially shows atrophy of dermal structures, whereas photoaged skin is predominately characterized by the accumulation of aberrant elastin fibers and GAGs, together with damaged and diminished collagen [61].

Conclusions

The face is the most exposed body site to environmental and external factors, being the physical area where the signs of skin aging initially appear and become more evident over time. Particularly, the periorbital and the perioral area are the two main sites involved in skin aging changes such as development of wrinkles, increased skin laxity, changes in skin pigmentation, etc.

We have just shown above the main examples of skin aging effects on these areas, highlighting also the numerous different mechanisms and the diverse skin structures involved. Particularly, since aging is always the result of a variable impact of extrinsic aging superimposed on the level of intrinsic aging, we have deeply described the multiple mechanisms of both intrinsic and extrinsic aging, highlighting also the fact that they may also show significant overlap (ROS production, DNA damage, telomere shortening, MMP overexpression, ECM degradation, etc.), possibly reflecting on shared clinical and histopathological aging alterations.

The knowledge of the complex skin aging mechanisms is of indisputable importance since it represents the only way to face efficaciously skin aging effects and consequently to be able to put the basis for the use of proper and safe intervention modalities.

References

1. American Society of Aesthetic Plastic Surgery (2006) 11.5 million cosmetic procedures in 2005. http://www.surgery.org/media/news-releases/115-million-cosmetic-procedures-in-2005. Accessed 02 Sep 2015
2. Baier J, Maisch T, Maier M et al (2006) Singlet oxygen generation by UVA light exposure of endogenous photosensitizers. Biophys J 91:1452–1459
3. Balzani A, Chilgar RM, Nicoli M et al (2013) Novel approach with fractional ultrapulse CO2 laser for the treatment of upper eyelid dermatochalasis and periorbital rejuvenation. Lasers Med Sci 28:1483–1487
4. Bernhard D, Moser C, Backovic A et al (2007) Cigarette smoke-an aging accelerator? Exp Gerontol 42:160–165
5. Birkedal-Hansen H (1987) Catabolism and turnover of collagens: collagenases. Methods Enzymol 144:140–171
6. Brandt FS, Cazzaniga A (2008) Hyaluronic acid gel fillers in the management of facial aging. Clin Interv Aging 3:153–159
7. Centers for Disease Control and Prevention (CDC) (2003) Trends in aging United States and worldwide. MMWR Morb Mortal Wkly Rep 52:101–104
8. Christensen K, Doblhammer G, Rau R et al (2009) Ageing populations: the challenges ahead. Lancet 374:1196–1208
9. Chung JH, Yano K, Lee MK et al (2002) Differential effects of photoaging vs intrinsic aging on the vascularization of human skin. Arch Dermatol 138:1437–1442
10. Chung JH, Eun HC (2007) Angiogenesis in skin aging and photoaging. J Dermatol 34:593–600
11. Coleman SR, Grover R (2006) The anatomy of the aging face: volume loss and changes in 3-dimensional topography. Aesthet Surg J 26:S4–S9
12. Dahmane R, Poljsak B (2011) Free radicals and intrinsic skin aging: basic principles. Health Med 5:1647–1654
13. Dai G, Freudenberger T, Zipper P et al (2007) Chronic ultraviolet B irradiation causes loss of hyaluronic acid from mouse dermis because of down-regulation of hyaluronic acid synthases. Am J Pathol 171:1451–1461
14. Davis EC, Callender VD (2011) Aesthetic dermatology for aging ethnic skin. Dermatol Surg 37:901–917
15. Desai S, Upadhyay M, Nanda R (2009) Dynamic smile analysis: changes with age. Am J Orthod Dentofacial Orthop 136:310–311
16. Dimri GP, Lee X, Basile G et al (1995) A biomarker that identifies senescent human cells in culture and in aging skin in vivo. Proc Natl Acad Sci U S A 92:9363–9367
17. Erbagci I, Erbagci H, Kizilkan N et al (1994) The effect of age and gender on the anatomic structure of Caucasian healthy eyelids. Am J Ophthalmol 117:231–234
18. Farage MA, Miller KW, Elsner P et al (2008) Intrinsic and extrinsic factors in skin ageing: a review. Int J Cosmet Sci 30:87–95
19. Fisher GJ, Wang ZQ, Datta SC et al (1997) Pathophysiology of premature skin aging induced by ultraviolet light. N Engl J Med 337:1419–1428
20. Fisher GJ, Kang S, Varani J et al (2002) Mechanisms of photoaging and chronological skin aging. Arch Dermatol 138:1462–1470

21. Friedman O (2005) Changes associated with the aging face. Facial Plast Surg Clin North Am 13:371–380
22. Furnas DW (1978) Festoons of orbicularis muscle as a cause of baggy eyelids. Plast Reconstr Surg 61:540–546
23. Furnas DW (1993) Festoons, mounds, and bags of the eyelids and cheek. Clin Plast Surg 20:367–385
24. Giardini MA, Segatto M, da Silva MS et al (2014) Telomere and telomerase biology. Prog Mol Biol Transl Sci 125:1–40
25. Gilchrest BA (1989) Skin aging and photoaging: an overview. J Am Acad Dermatol 21:610–613
26. Gilchrest BA (1996) A review of skin ageing and its medical therapy. Br J Dermatol 135:867–875
27. Gilchrest BA (2013) Photoaging. J Invest Dermatol 133:E2–E6
28. Goh C (2009) The need for evidence-based aesthetic dermatology practice. J Cutan Aesthet Surg 2:65–71
29. Gonzalez-Ulloa M, Flores ES (1965) Senility of the face: basic study to understand its causes and effects. Plast Reconstr Surg 36:239–246
30. Goodsell DS (2001) The molecular perspective: ultraviolet light and pyrimidine dimers. Stem Cells 19:348–349
31. Grozdev IS, Van Voorhees AS, Gottlieb AB et al (2011) Psoriasis in the elderly: from the Medical Board of the National Psoriasis Foundation. J Am Acad Dermatol 65:537–545
32. Harley CB, Futcher AB, Greider CW (1990) Telomeres shorten during ageing of human fibroblasts. Nature 345:458–460
33. Heck DE, Vetrano AM, Mariano TM et al (2003) UVB light stimulates production of reactive oxygen species: unexpected role for catalase. J Biol Chem 278:22432–22436
34. Herrmann G, Wlaschek M, Lange TS et al (1993) UVA irradiation stimulates the synthesis of various matrix-metalloproteinases (MMPs) in cultured human fibroblasts. Exp Dermatol 2:92–97
35. Hussein MR (2005) Ultraviolet radiation and skin cancer: molecular mechanisms. J Cutan Pathol 32:191–205
36. Ichihashi M, Fujiwara Y (1981) Clinical and photobiological characteristics of Japanese xeroderma pigmentosum variant. Br J Dermatol 105:1–12
37. Ichihashi M, Ueda M, Budiyanto A et al (2000) UV-induced skin damage. Toxicology 189:21–39
38. Inomata S, Matsunaga Y, Amano S et al (2003) Possible involvement of gelatinases in basement membrane damage and wrinkle formation in chronically ultraviolet B-exposed hairless mouse. J Invest Dermatol 120:128–134
39. Iovine B, Nino M, Irace C et al (2009) Ultraviolet B and A irradiation induces fibromodulin expression in human fibroblasts in vitro. Biochimie 91:364–372
40. Jenkins G (2002) Molecular mechanisms of skin ageing. Mech Ageing Dev 123:801–810
41. Kammeyer A, Luiten RM (2015) Oxidation events and skin aging. Ageing Res Rev 21:16–29
42. Kang S, Fisher GJ, Voorhees JJ (2001) Photoaging: pathogenesis, prevention, and treatment. Clin Geriatr Med 17:643–659
43. Kjellén L, Lindahl U (1991) Proteoglycans: structures and interactions. Annu Rev Biochem 60:443–475
44. Kondo S (2000) The roles of cytokines in photoaging. J Dermatol Sci 23:S30–S36
45. Kurban RS, Bhawan J (1990) Histologic changes in skin associated with aging. J Dermatol Surg Oncol 16:908–914
46. Landau M (2007) Exogenous factors in skin aging. Curr Prob Dermatol 35:1–13
47. Lee DE, Chung MY, Lim TG et al (2013) Quercetin suppresses intracellular ROS formation, MMP activation, and cell motility in human fibrosarcoma cells. J Food Sci 78:1464–1469
48. Li T, Ning K, Le J et al (2012) Tumor suppression in the absence of p53-mediated cell-cycle arrest, apoptosis, and senescence. Cell 149:1269–1283
49. Love LP, Farrior EH (2010) Periocular anatomy and aging. Facial Plast Surg Clin North Am 18:411–417
50. Mera SL, Lovell CR, Jones RR et al (1987) Elastic fibres in normal and sun-damaged skin: an immunohistochemical study. Br J Dermatol 117:21–27
51. Millis AJ, Hoyle M, McCue HM et al (1992) Differential expression of met- alloproteinase and tissue inhibitor of metalloproteinase genes in aged human fibroblasts. Exp Cell Res 201:373–379
52. Muto J, Kuroda K, Wachi H et al (2007) Accumulation of elafin in actinic elastosis of sun-damaged skin: elafin binds to elastin and prevents elastolytic degradation. J Invest Dermatol 127:1358–1366
53. Naylor EC, Watson RE, Sherratt MJ (2011) Molecular aspects of skin ageing. Maturitas 69:249–256
54. Ortonne JP (2002) Photoprotective properties of skin melanin. Br J Dermatol 146:7–10
55. Oxlund H, Andreassen TT (1980) The roles of hyaluronic acid, collagen and elastin in the mechanical properties of connective tissues. J Anat 131: 611–620
56. Perkins SW, Sandel HD 4th (2007) Anatomic considerations, analysis, and the aging process of the perioral region. Facial Plast Surg Clin North Am 15:403–407
57. Pfeifer GP, You YH, Besaratinia A (2005) Mutations induced by ultraviolet light. Mutat Res 571:19–31
58. Poljsak B, Dahmane RG, Godic A (2012) Intrinsic skin aging: the role of oxidative stress. Acta Dermatovenerol Alp Panon Adriat 21:33–36
59. Pons B, Belmont AS, Masson-Genteuil G et al (2010) Age-associated modifications of Base Excision Repair activities in human skin fibroblast extracts. Mech Ageing Dev 131:661–665
60. Rinnerthaler M, Bischof J, Streubel MK et al (2015) Oxidative stress in aging human skin. Biomolecules 5:545–589
61. Scharffetter-Kochanek K, Brenneisen P, Wenk J et al (2000) Photoaging of the skin from phenotype to mechanisms. Exp Gerontol 35:307–316

62. Shapiro DP (1999) Geriatric demographics and the practice of otolaryngology. Ear Nose Throat J 78: 418–421
63. Situm M, Buljan M, Cavka V et al (2010) Skin changes in the elderly people-how strong is the influence of the UV radiation on skin aging? Coll Antropol 34:9–13
64. Sjerobabski-Masnec I, Situm M (2010) Skin aging. Acta Clin Croat 49:515–518
65. Starcher B, Conrad M (1995) A role for neutrophil elastase in solar elastosis. Ciba Found Symp 192: 338–346
66. Truswell WH 4th (2013) Aging changes of the periorbita, cheeks, and midface. Facial Plast Surg 29:3–12
67. Tsutsumi M, Denda M (2007) Paradoxical effects of beta-estradiol on epidermal permeability barrier homeostasis. Br J Dermatol 157:776–779
68. Varani J, Schuger L, Dame MK et al (2004) Reduced fibroblast interaction with intact collagen as a mechanism for depressed collagen synthesis in photodamaged skin. J Invest Dermatol 122:1471–1479
69. Warren R, Gartstein V, Kligman AM et al (1991) Age, sunlight, and facial skin: a histologic and quantitative study. J Am Acad Dermatol 25:751–760
70. Wlaschek M, Tantcheva-Poor I, Naderi L et al (2001) Solar UV irradiation and dermal photoaging. J Photochem Photobiol 63:41–51
71. Wollina U (2013) Perioral rejuvenation: restoration of attractiveness in aging females by minimally invasive procedures. Clin Interv Aging 8:1149–1155
72. Woodley DT, Kalebec T, Banes AJ et al (1986) Adult human keratinocytes migrating over nonviable dermal collagen produce collagenolytic enzymes that degrade type I and type IV collagen. J Invest Dermatol 86:418–423
73. Yaar M, Gilchrest BA (2012) Aging of the skin. In: Goldsmith LA, Katz SI, Gilchrest BA et al (eds) Fitzpatrick's dermatology in general medicine. McGra Hill, New York, pp 1213–1226
74. Zouboulis CC, Chen WC, Thornton MJ et al (2007) Sexual hormones in human skin. Horm Metab Res 39:85–95
75. Zouboulis CC, Makrantonaki E (2011) Clinical aspects and molecular diagnostics of skin aging. Clin Dermatol 29:3–14

Aesthetic Procedures for Increased Lip Volume: Hyaluronic Acid Fillers in Nonsurgical Lip and Eye Rejuvenation Techniques

3

Giselle Prado, Sonal Choudhary, and Martin Zaiac

3.1 Introduction

Patients often present with many reasons for wanting to obtain lip augmentation or restoration; invariably, the majority complains of an aging look or thinning to the perioral region and requests lip reshaping due to loss of volume. The commonly encountered signs of perioral aging include decreased vermillion showing, blunting of the Cupid's bow, less visible white roll, vertical rhytids, marionette lines, formation of a mentalis crease, and deep nasolabial folds (Fig. 3.1) [1]. Younger patients present for lip augmentation in order to obtain the ever-changing look that is desirable at the moment.

It appears that atrophy and not descent accounts for most of the aging of the face [2]. There is a loss of glycosaminoglycans and proteoglycans in the dermis and a simultaneous decrease in the amount of collagen and elastic fibers. The repeated contraction of facial muscles also leads to rhytid formation in the perioral region [3]. On the macrolevel as a person ages, the lips become flatter, the upper lip elongates, the Cupid's bow is lost, and the oral commissures descend [4]. Resorption of the facial bones with ensuing soft tissue repositioning has also been implicated as a cause of aging in the perioral region [5]. Injection of hyaluronic acid, a glycosaminoglycan, acts to increase the water content of the lips for a plumper, fuller appearance [3]. Normal lip aging can be exacerbated by sun exposure and smoking [4].

Hyaluronic acid (HA) is a commonly used temporary filler for lip augmentation due to its hygroscopic properties and resulting natural appearance. The effects last between 3 and 6 months. Some patients may experience long-term results from HA fillers due to neocollagenesis induced by mechanical tension on fibroblasts from HA injection [6]. It is the author's opinion that the trauma induced by injection leads to an inflammatory process that also extends the effects of HA fillers. They do not require skin testing unlike the collagen-based fillers. HA fillers consist of an uncross-linked soluble HA phase and an insoluble cross-linked fraction of HA particles of a predetermined size [2]. HA is cross-linked in order to

G. Prado, BS
Department of Dermatology, Herbert Wertheim College of Medicine, Florida International University, Miami, FL, USA
e-mail: gprad009@fiu.edu

S. Choudhary, MD
Department of Dermatology, University of Miami Miller School of Medicine, Miami, FL, USA
e-mail: drsonalchoudhary@gmail.com

M. Zaiac, MD (✉)
Department of Dermatology, Herbert Wertheim College of Medicine, Florida International University, Miami, FL, USA

Greater Miami Skin and Laser Center, Miami, FL, USA
e-mail: drmartyz@aol.com

G. Fabbrocini et al. (eds.), *Nonsurgical Lip and Eye Rejuvenation Techniques*,
DOI 10.1007/978-3-319-23270-6_3

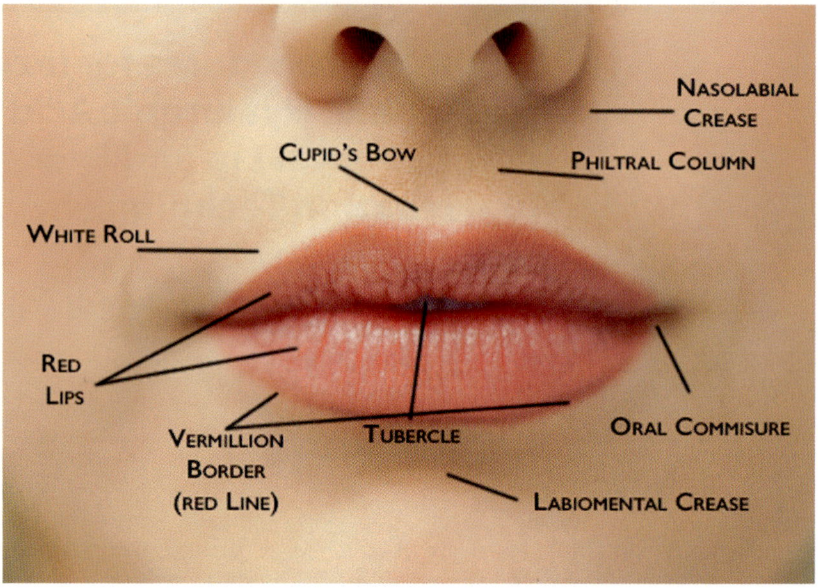

Fig. 3.1 Characteristic changes of aging to the perioral area include flattening of the Cupid's bow, perioral rhytids, blunting of the vermilion border, and descent of the oral commissures

increase the half-life after injection [2]. The degree of cross-linking within the HA filler corresponds to the firmness of the resulting gel. The firmness of HA fillers is measured by the elastic modulus (G'), and higher numbers correspond to firmer products (Restylane > Juvederm Ultra > Belotero Balance) [2]. Physicians must keep in mind the firmness of the product they are injecting in order to achieve natural feeling lips [7]. They have desirable safety profiles due to the reversibility by enzymatic degradation using hyalurodinase.

Currently, there are two FDA-approved HA fillers specifically for lip augmentation: Restylane Injectable Gel and Restylane Silk (Galderma S.A., Lausanne, Switzerland) [8]. However, off-label uses of Juvederm Ultra (Allergan, Inc., Irvine, CA), Captique (Inamed Corporation, Santa Barbara, CA), and Belotero Balance (Merz North America, Greensboro, NC) include lip augmentation [9]. Restylane, Captique, and Juvederm are bacterially derived from *Streptococcus equi* and thus have low risk of immunogenicity and allergic reaction.

Restylane is a cross-linked HA gel that is classified as a nonanimal stabilized HA filler. It can be used to treat nasolabial folds, marionette lines, and glabellar lines and to augment the lips. Side effects of injection with Restylane can last up to 1 week and include redness, swelling, bruising, and induration [10]. It is firm filler that does not spread out after injection because of its viscosity and higher elastic modulus.

Captique replaced Hylaform and Hylaform Plus from the Inamed Corporation seeing as they are no longer used [2]. Captique is also classified as a nonanimal stabilized HA.

Juvederm Ultra (classified as Hylacross HA) is a less viscous long-chain HA gel with a lower elastic modulus that tends to spread more after injection [2]. It is also more concentrated and hydrophilic than Restylane and thus will absorb more water from the surrounding tissues [11].

Belotero Balance (classified as cohesive polydensified matrix HA) is the newest HA filler to come to market and has the lowest viscosity and elastic modulus between Restylane and Juvederm [2]. It has the greatest capability to spread after injection, which is desirable when performing lip augmentation in order to preserve pliability of the lips [2]. It's especially effective as a superficial filler when used to restore areas of vertical rhytids to a more youthful appearance.

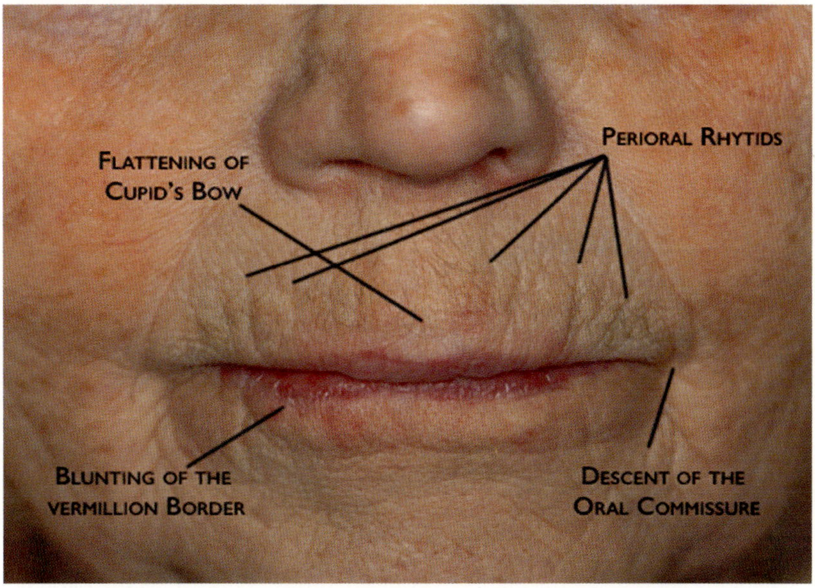

Fig. 3.2 Important anatomic landmarks of the perioral area for lip augmentation

3.2 Guidelines

When preparing to inject a patient with any filler, the physician should bear in mind the important anatomy surrounding the perioral region (Fig. 3.2). The lip is mainly composed of the orbicularis oris muscle with an overlying fascia. However, many muscles insert into the angle of the mouth and should be considered when injecting. The corresponding vermillion or cutaneous layers sit on top of the fascia [1].

The vascular supply of the perioral region consists of the facial artery and its direct branch into the lip, the labial artery. The infraorbital and mental nerves provide this area with sensation and motor function [3].

The operator must keep in mind the aesthetics of lip augmentation, which is subject to constantly changing social pressure, in order to achieve high patient satisfaction and cosmetically acceptable results. The lower lip should be fuller with more vermillion show than the upper lip [4]. In Caucasian women, it is advisable to follow the golden ratio of 1:1.618 in terms of volume in the upper and lower lip. Black and Asian women may have proportions that approximate 1:1 [12]. When viewing the patient in profile, the upper lip should project slightly farther than the lower lip. The white roll separates the facial skin from the vermillion of the lip. It must be continuous and smooth because a misalignment as small as 1 mm can be noticeable to observers [4]. The Glogau-Klein point (the elevation or ski slope in the upper lip where the skin turns into red vermillion at the arches) can be enhanced to project the upper lip and create the ski slope to look characteristic of younger lips.

There are natural prominences in the upper and lower lips that must be maintained to achieve the pouty look. These include the two tubercles that lie lateral to the midline on both sides of the lower lip, the tubercle that lies in the midline of the upper lip, and the two tubercles that lie in the lateral corners of the upper lip [9].

When performing lip augmentation, it is essential that the doctor appropriately fill lips to satisfy patient expectations but not overfill them so as to create a "duckbill" appearance. It is preferable to treat again later if the desired cosmetic effect is not achieved [10].

In order for the patient to observe the changing lip proportions, it is useful to have a hand mirror available and pre- and post-procedure photographs taken from the anterior and lateral view.

Swelling will not occur symmetrically after injection so the injector should treat each side of the lip for each step before moving onto the next step [9].

Additionally, injectors need to be cognizant of the amount of volume that has been used at each step of the procedure in order to avoid the unwanted consequence of running short of product before completing the augmentation. In patients who only agree to pay for a certain amount of product, this amount must guide and limit the extent of augmentation.

Viscous fillers such as Restylane should be injected more deeply to avoid visible nodules and the Tyndall effect [13]. Less viscous fillers such as Juvederm and Belotero can be used more superficially to achieve natural results.

Patients should be educated and counseled on realistic expectations of results. Visual scales are used to grade lip augmentation pre- and post-procedure such as the lip fullness scale [14].

3.3 Author's Technique

Prior to any lip augmentation procedure, it is essential to understand the importance of facial proportions and anatomy. While some studies have shown that facial symmetry is an important factor to define a more desirable face, it has also been shown that for a given individual there were statistically significant lower ratings of attractiveness for perfectly symmetrical computer-generated left-left and right-right faces compared to natural faces [15]. Clearly, the goal of lip augmentation should not be complete symmetry. Natural asymmetry makes a face and its expressions unique and attractive. Injectors should bear this in mind when performing augmentation of facial features and exercise the best scientific and artistic skills that will lead to aesthetically pleasing results.

Firstly, it is vital to gauge the expectations of the patient prior to beginning injection. Patients should be counseled to expect subtle improvement of their facial appearance with the help of lip augmentation within the boundaries of their individual natural features. Patients presenting with a desire to completely change their appearance or to change their lips to resemble a celebrity's have unrealistic goals. This should be discussed and clarified at the first visit. Appropriate expectations help achieve good outcomes and satisfied patients.

Following a discussion on expectations, patient should be provided with a hand mirror to allow them to point to areas of concern. This helps the clinician understand the exact areas that need augmentation in the patient's mind. For example, a patient may suggest that her upper lip is too thin in the middle or the corners of the mouth are downturned. Doing this exercise with the patient helps build good rapport, puts the doctor on the same page as the patient, and prevents dissatisfaction after the procedure. The doctor should pay special attention to the elements of concern as the patient sees it.

Photographs must be taken from the front and side profiles under standard lighting and background before and after procedures (Figs. 3.3, 3.4, 3.5, and 3.6). This allows for appropriate follow-up, understanding and managing complications, guiding future treatments, and providing documentation.

Once the patient is ready for the procedure, topical anesthesia is used to anesthetize the treatment site. This author prefers topical anesthesia instead of nerve blocks as the former allows for maintenance of lip movement during injection, thus helping to assess the amount of product injected and its effect on appearance. Nerve blocks cause loss of lip movement and affect real-time assessment of results. The anesthetic of choice is 30 % lidocaine in petrolatum ointment for 30–40 min pre-procedure.

If the treatment involves the nasolabial folds, cheeks, and/or upper cutaneous lip, these should be augmented first followed by the lips, seeing as the lidocaine will spread from these areas toward the infraorbital nerve and help with anesthetizing the lips.

Once the area is anesthetized, the patient is seated on the treatment chair in a partially inclined, well-supported position, comfortable to the patient and the doctor. This is necessary for a slow, well-controlled injection.

The needle is primed with a bolus of the product. Anterograde injection with this drop helps to

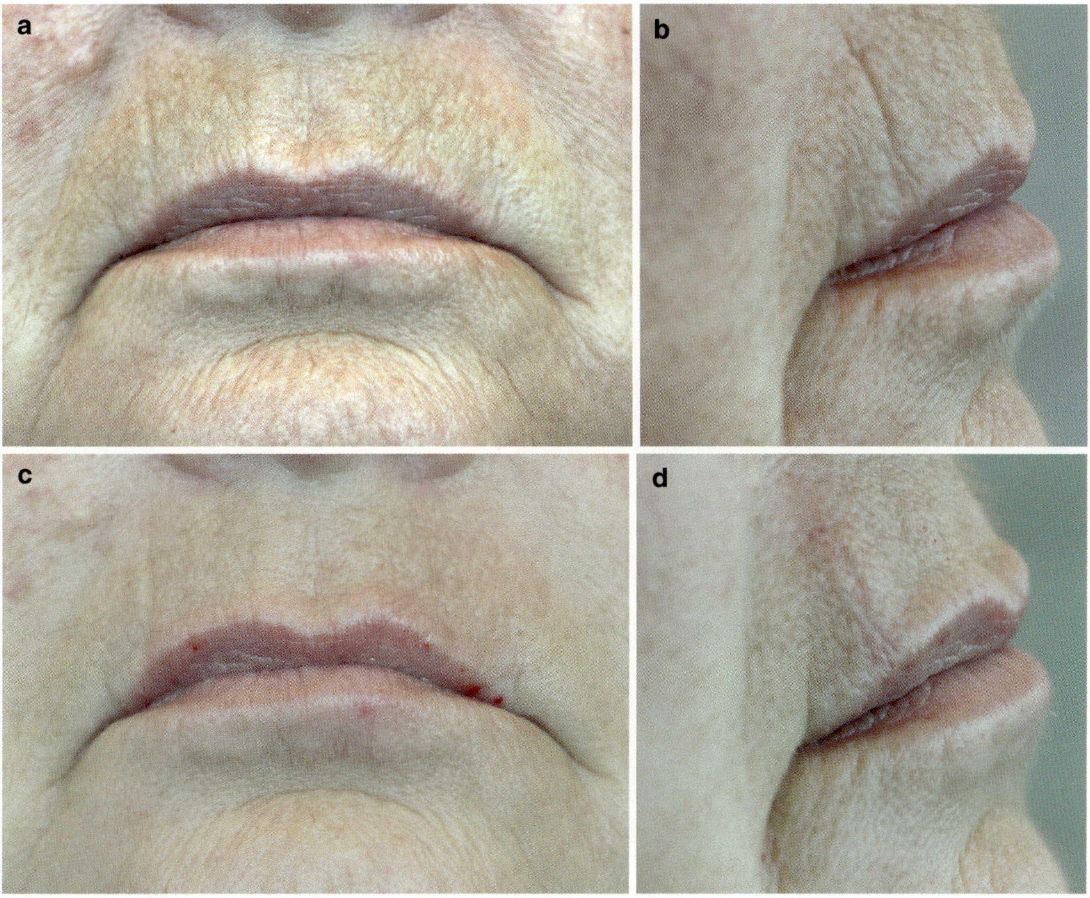

Fig. 3.3 (**a**, **b**) Older woman before lip augmentation shown in frontal and lateral views. (**c**, **d**) Same woman immediately post-lip injection of hyaluronic acid filler shown in frontal and lateral views. Note improvement of perioral rhytids

roll any vasculature out of the way of the needle instead of puncturing through them.

The eyesight and tip of needle are in one line with the lip for clear view of the injection and its immediate effect.

To enhance the vermillion border, the needle is inserted at the initial point such that it enters parallel into the natural tunnellike space between the white roll and vermillion with constant anterograde pressure (Fig. 3.3). The initial point will depend on the plan of augmentation for an individual patient and on their natural lip architecture. Some patients may require a more lateral starting point while others can be started more medially. The tunnel-like space is between the wet and dry sides of the lip. It is important to deliver augmentation into the correct anatomical region to gain optimal aesthetic effects and avoid poor outcomes.

Injection is continued slowly with small volumes injected with steady pressure for even results.

If the overall lip volume is being augmented, this author's preferred method is to inject from the vermillion border with the needle angled parallel into the red pulp. This leads to accentuation of the tubercles and less pain than experienced when injecting directly into the red pulp.

Alternatively, the injections can be continued in the pulp area of the lip in the same manner as above, layered underneath the first layer and using it as a support. For a more pouty appearance (Paris lip), dermal filler is first placed along the border of the upper lip, adding definition. Then

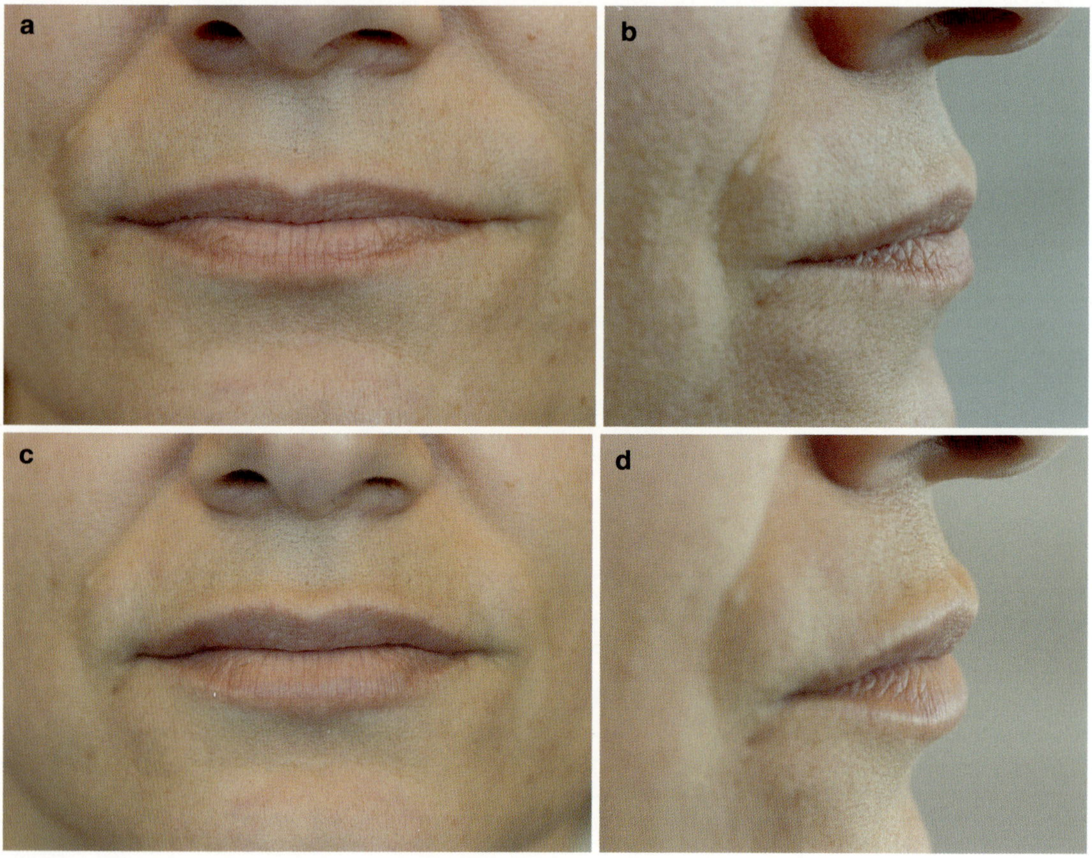

Fig. 3.4 (**a**, **b**) Middle-aged woman before lip augmentation shown in frontal and lateral views. (**c**, **d**) Middle-aged woman after injection of hyaluronic acid filler shown in frontal and lateral views. Note increased plumpness and heightened upper lip

the Cupid's bow is accentuated to create a distinctive "V" shape. Afterward, the vertical arches between the upper lip and nose are built up. This creates the distinctive peaks that are the signature of the Paris lip. Finally, when augmenting the lower lip, focus on the middle third as the area of injection. If you augment the entire lower lip, a rounded, duck-like appearance will result.

If the patient has a previous scar in the area of injection, the technique should be modified such that the scar area is bypassed. The scar prevents proper flow of the product across the scar tissue.

In certain patients, onabotulinum toxin may be injected into the orbicularis oris in combination with lip fillers to produce elevation of the lip along with augmentation and reduction in perioral rhytids.

3.3.1 Post-procedure Care

To prevent bruising, light pressure is applied to the injection sites. The pressure should not be firm as this can cause displacement of the filler product. Patient is also advised to hold an ice pack on the area for 15–20 min or longer if on a blood-thinning medication.

Patients should be instructed that HA absorbs water and causes some swelling in addition to that caused by the trauma of the injection itself. This edema should subside by 10–15 % in 48–72 h. If bruising is noted on follow-up, pulsed dye laser may be used 48 h post-procedure to lighten the color of the bruise and speed up the healing process.

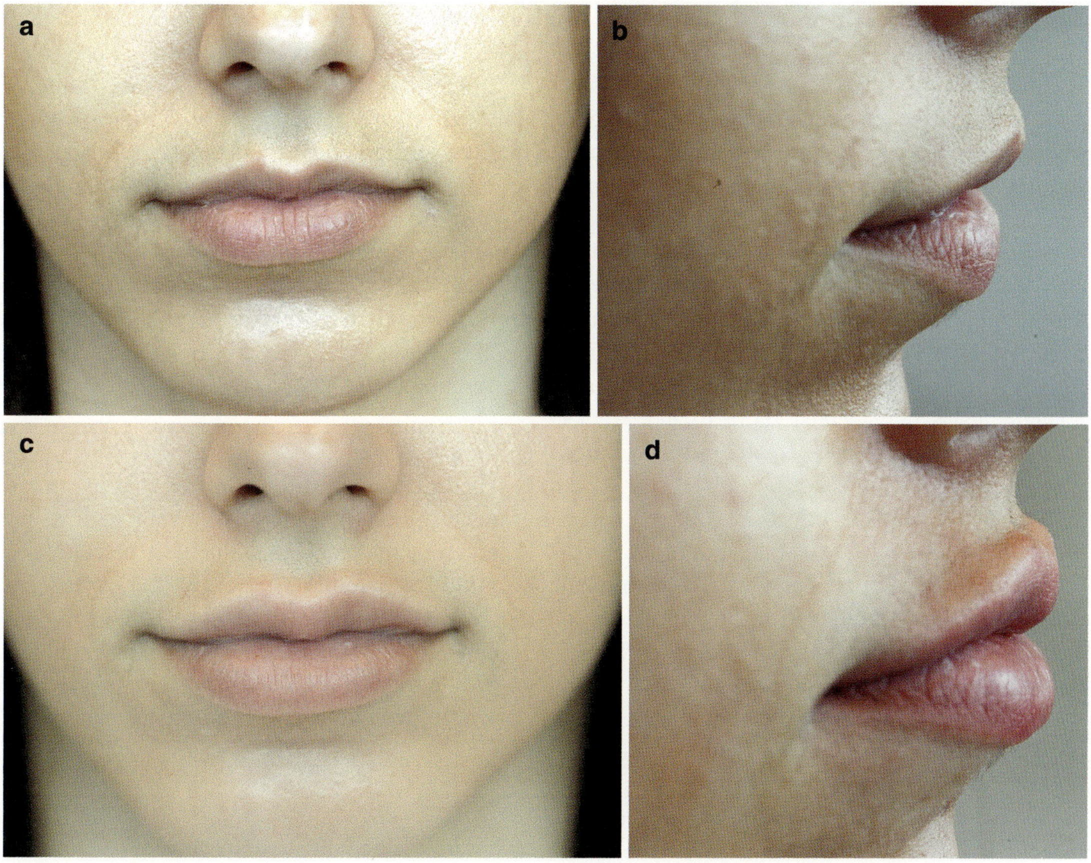

Fig. 3.5 (**a**, **b**) Young woman before lip augmentation shown in frontal and lateral views. (**c**, **d**) Young woman after injection of hyaluronic acid filler shown in frontal and lateral views. Note fuller upper and lower lips

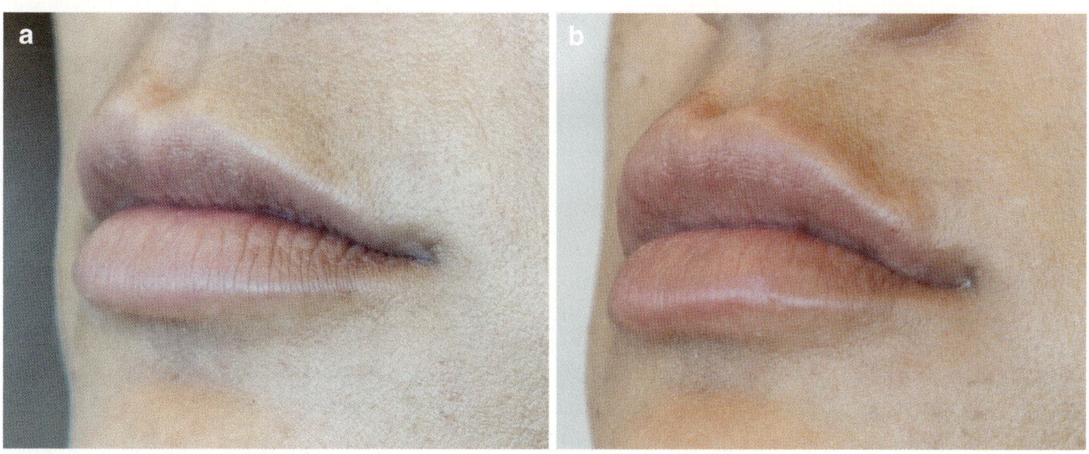

Fig. 3.6 Patient shown on oblique view before (**a**) and after (**b**) lip augmentation

3.4 Complications

Hyaluronic acid fillers are well tolerated and rarely result in adverse reactions for patients due to their temporary nature and low immunogenicity [16, 17]. Commonly experienced reactions due to injection include local inflammation, hyperemia, bruising, redness, warmth, tenderness, and ecchymosis [16–18]. Hypersensitivity and an angioedematous swelling can be seen in some patients [16, 17]. Bruising and swelling usually abate within 7 days [16]. Erythema and telangiectasias may persist more than 2 weeks after injection and can be treated with hyalurodinase [16].

Infection from filler injection is also a cause for concern and can present as induration, erythema, itching, and tenderness. The clinician should use an alcohol swab to prepare a sterile site for the injection, thus avoiding the risk of transdermally acquired bacterial infection from the presence of biofilms [17]. Patients with a history of herpes simplex labialis can have reactivation of infection after injection and thus should be given prophylaxis with antiviral medications (acyclovir or valacyclovir) 2 days before their procedure and for 5 days after [17]. Painful, red nodules that present within 3–14 days after injection are suspicious for infection and should be cultured. Antibiotic treatment may be indicated depending on culture results, and steroids may be added depending on the amount of inflammation [16]. It is not recommended that injectors store leftover product due to increased infection risk.

Some complications arise from the inherent technique and expertise of the operator. The Tyndall effect can occur with certain HA fillers when the filler is injected too superficially, and thus a bluish discoloration results at the injection site [16]. Vascular occlusion of the blood supply of the lips can result when injecting directly into an artery or vein or from compression. The operator should take care to avoid the superior and inferior labial arteries when injecting the lips. Hyalurodinase should be immediately injected locally with flooding of the area if severe localized pain or unusual blanching is noted. Firm massage to the area is also indicated [18]. Necrosis of the occluded area can occur if this is not recognized and treated promptly [17].

Immune-related adverse events include delayed-type hypersensitivity and the development of foreign body granulomas. Foreign body granulomatous reactions are rare when using hyaluronic acid but lead to the presence of lumps or small nodules on the lips [19]. Differentiating between an immune process and a normal reaction has largely been based on the intensity of the reaction and clinician preference [16].

Conclusion

Hyaluronic acid fillers are an effective treatment option for patients who want to augment and revitalize their lips. Injectors must keep in mind the natural aesthetics of the patient and patient preferences when providing this procedure. Major adverse effects are rare, and thus HA fillers are a first-line treatment modality for the nonpermanent augmentation of lips.

References

1. Ali MJ, Ende K, Maas CS (2007) Perioral rejuvenation and lip augmentation. Facial Plast Surg Clin North Am 15(4):491–500, vii
2. Sundaram H, Cassuto D (2013) Biophysical characteristics of hyaluronic acid soft-tissue fillers and their relevance to aesthetic applications. Plast Reconstr Surg 132(4 Suppl 2):5S–21S
3. Hotta T (2006) Understanding the anatomy when using dermal fillers enhances patient safety. Plast Surg Nurs 26(3):149–151
4. Perkins SW, Sandel HD (2007) Anatomic considerations, analysis, and the aging process of the perioral region. Facial Plast Surg Clin North Am 15(4):403–407, v
5. Vleggaar D, Fitzgerald R (2008) Dermatological implications of skeletal aging: a focus on supraperiosteal volumization for perioral rejuvenation. J Drugs Dermatol 7(3):209–220
6. Koger C, Cohen J (2014) The lasting effects of fillers through neocollagenesis. Dermatologist 22(4)
7. Cartier H, Trevidic P, Rzany B et al (2012) Perioral rejuvenation with a range of customized hyaluronic acid fillers: efficacy and safety over six months with a specific focus on the lips. J Drugs Dermatol 11(1 Suppl):s17–s26

8. Soft Tissue Fillers Approved by the Center for Devices and Radiological Health. U.S. Food and Drug Administration. Accessed October 29, 2014.
9. Sarnoff DS, Gotkin RH (2012) Six steps to the "perfect" lip. J Drugs Dermatol 11(9):1081–1088
10. Monheit GD, Coleman KM (2006) Hyaluronic acid fillers. Dermatol Ther 19(3):141–150
11. Eccleston D, Murphy DK (2012) Juvéderm(®) Volbella™ in the perioral area: a 12-month prospective, multicenter, open-label study. Clin Cosmet Investig Dermatol 5:167–172
12. Goodman G (2012) Duckless lips: how to rejuvenate the older lip naturally and appropriately. Cosmetic Dermatol 25(6):276
13. Monheit GD (2007) Hyaluronic acid fillers: Hylaform and Captique. Facial Plast Surg Clin North Am 15(1): 77–84, vii
14. Carruthers A, Carruthers J, Hardas B et al (2008) A validated lip fullness grading scale. Dermatol Surg 34(Suppl 2):S161–S166
15. Zaidel DW, Deblieck C (2007) Attractiveness of natural faces compared to computer constructed perfectly symmetrical faces. Int J Neurosci 117(4):423–431
16. Cox SE, Adigun CG (2011) Complications of injectable fillers and neurotoxins. Dermatol Ther 24(6):524–536
17. Gilbert E, Hui A, Meehan S, Waldorf HA (2012) The basic science of dermal fillers: past and present part II: adverse effects. J Drugs Dermatol 11(9): 1069–1077
18. Vent J, Lefarth F, Massing T, Angerstein W (2014) Do you know where your fillers go? An ultrastructural investigation of the lips. Clin Cosmet Investig Dermatol 7:191–199
19. Eversole R, Tran K, Hansen D, Campbell J (2013) Lip augmentation dermal filler reactions, histopathologic features. Head Neck Pathol 7(3):241–249

Aesthetic Procedures for Lip Wrinkles: Skin Needling and Botox

4

Gabriella Fabbrocini and Luigia Panariello

Lip wrinkles are fine or deep lines that can be observed around the mouth. They appear as vertical lip lines perpendicular to the vermillion border and can be divided in *static* and *dynamic wrinkles*.

Static wrinkling can be caused by several factors, such as age, sun exposure, cigarette smoking, as well as unknown causes like genetics, gender differences, and intrinsic soft tissue characteristics.

Dynamic perioral wrinkles are caused by muscle contractions, which can be voluntary (e.g., smoking or playing wind instruments) or involuntary (e.g., smiling or grimacing).

Many aesthetic procedures for lip wrinkles are available: *static wrinkles* can be treated through facial skin resurfacing, laser, mechanical dermabrasion, skin needling, chemical peels, and soft tissue fillers; for *dynamic wrinkles*, BOTOX® injections can be very useful.

This wide selection of techniques allows us to choose those with higher efficacy, minimal adverse effects, and quick healing time. In particular skin needling and BOTOX® injections are the newest procedures that share all the advantages listed above.

G. Fabbrocini (✉) • L. Panariello
Section of Dermatology, Department of Clinical Medicine and Surgery, University of Naples Federico II, Via Pansini 5, Naples 80131, Italy
e-mail: gafabbro@unina.it

4.1 Anatomy Elements

A good knowledge of perioral region's anatomy is essential for a careful approach to aesthetic procedures, in particular to provide the best treatment options for each patient and to avoid side effects.

The perioral muscles are arranged in interlacing and decussating bundles organized in several layers.

They can be classified into three groups based on insertion.

- Group I muscles insert into the modiolus; they are orbicularis oris, buccinator, levator anguli oris, depressor anguli oris, zygomaticus major, and risorius.
- Group II muscles insert into the upper lip; they are levator labii superioris, levator labii superioris alaeque nasi, and zygomaticus minor.
- Group III muscles insert into the lower lip; they are depressor labii inferioris, mentalis, and platysma.

4.2 Skin Needling

Skin needling, also called dermarolling, percutaneous collagen induction (PCI), or collagen induction, is an efficient technique used since 1995 in many aesthetic procedures [1, 2].

G. Fabbrocini et al. (eds.), *Nonsurgical Lip and Eye Rejuvenation Techniques*,
DOI 10.1007/978-3-319-23270-6_4

To date, skin needling, alone or combined with other methods, has been proposed as an effective method to treat scars and wrinkles, but also to enhance penetration of topical substances in dermatologic pathology such as melasma [3, 4].

The device used is a drum-shaped roller studded with a variable number of fine microneedles, which penetrate a depth of 0.5–3 mm into the skin creating thousands of micro-wounds without damaging the epidermidis.

The micro-holes, in fact, are about four cells in diameter so that the stratum corneum remains intact [5].

The needles must be rolled in four directions: vertically, horizontally, diagonally right to the left, and diagonally left to the right. Needling should be performed maintaining a mild pressure over the affected area, for a total of 10–14 passes [4].

Skin needling is effective in minimizing scars (acne scars, burn scars) and wrinkles and by promoting the neocollagenogenesis and the wound healing process through a cascade of growth factors [6].

Despite ablative laser treatment, skin needling does not cause thermal damage, and no signs of dermabrasive reduction of epidermal thickness are evident 24 h after the procedure so that neither increased nor decreased the number of melanocytes [7].

For these reasons, it is a procedure suitable for all skin types, it can be repeated safely, and it is also applicable in some skin areas that are not suitable for laser treatments and deep peels [8].

Skin needling represents a useful tool for the treatment of perioral wrinkles, especially for vertical lines, also called "bar code," and for perioral wrinkles, typical of smokers; it can also be an adjuvant treatment, in association with other techniques (such as filler injection), for the "marionette" wrinkles.

With repeated sessions, new collagen is gradually produced, plumping and filling, in a physiologic way, wrinkles, lines, and depressed scars.

4.2.1 Advantages of Skin Needling

- It is a minimally invasive procedure with rapid healing and little downtime.
- The epidermis is not damaged, so that it can be repeated.
- Risks of postinflammatory hypo- or hyperpigmentation are minimal.
- It is suitable for all skin types, even thin and previously lasered skin.
- It can be performed on all areas of the face, neck, and body.
- It costs less than laser treatments.

4.3 BOTOX

Botulinum toxin injection for treatment of facial wrinkles has been one of the most frequently performed aesthetic treatments in these last years.

Since botulism was first described in the eighteenth century, the botulinum toxin type A has been used first for skeletal muscle hyperactivity, and then it was being investigated for the treatment of many other conditions, until its approbation for temporary improvement of moderate to severe glabellar lines. It was approved for human use under the name of Oculinum™ first and then after acquisition by Allergan, Inc., under the name of BOTOX®.

Botulinum toxin type A acts by inhibiting the exocytosis of acetylcholine on cholinergic nerve endings of motor nerves. In fact after binding to the membrane of the nerve terminal, it is translocated into the neuronal cytosol where it cleaves one or more of the SNARE (soluble N-ethylmaleimide-sensitive factor attachment protein receptor) proteins. Since these proteins are necessary for vesicle docking and fusion, botulinum toxin type A reduces neurotransmitter release. This induced weakness of the muscle starts after 3–7 days and lasts for a period of 3–4 months [9, 10].

Botulinum toxin type A is most often injected into muscles that are overactive.

It results in decreased muscle activity.

For cosmetic use, the target of BOTOX® is the muscles of facial expression. In fact, with repeated contraction of these muscles, the skin overlying the muscle can develop wrinkles.

It is very important to distinguish dynamic wrinkles, caused by muscle contraction, from static wrinkles, caused by photo- and chrono-aging, because the latter is not affected by treatment with BOTOX®.

To distinguish them, physician can ask patients to grimace.

Botox can be used, for example, to correct glabella wrinkles, frown lines in the forehead, periorbital lines, hyperactivity in the muscles of the upper lip, and hypertrophy of the musculus masseter.

Many studies show that efficacy of botulinum toxin type A after several treatment sessions persists for many years without decrements in safety and without necessity of increased doses [11].

We can identify three important muscles to treat in the perioral region.

4.3.1 The Orbicularis Oris

Its function is to close the mouth by approximating the lips, to bring the lips together against the alveolar arch, and to protrude the lips. For these reasons, its hyperactivity is in part responsible for perioral wrinkle formation.

Because of its circular shape, it is recommended to treat both the upper and lower portions to maintain balance.

Injections must be superficial, performing it in the lower dermis and no deeper than the dermo-subcutaneous junction [12].

The dilution of BOTOX® should be 100 U vial of BOTOX® with anywhere from 1 to 4 ml of normal saline. In this way, it can spread over the superficial fibers of the orbicularis oris, treating the multitude of vertical lines across the lips.

In this way we can use only 1 or 2 U for each quadrant, with a total treatment dose of 4–6 U for the whole area [13].

Injections can be performed either into the border between the pars peripheralis and pars marginalis or 3–5 mm above the vermillion border into the lateral pars of the orbicularis oris, at least 1 cm from the mouth corner and avoiding the philtrum column area, for risk of flattening its lateral edges [12, 13].

The needle should be injected parallel to the skin surface.

4.3.2 Depressor Anguli Oris

Its function is to pull the corners of the mouth downward, moving the marionette lines down [14].

Injections should be superficial in the lower third of this muscle, with the needle directed laterally.

The dilution of BOTOX® should be 100 U vial of BOTOX® with 1 ml of normal saline. In this way we can avoid not only aesthetic adverse effects such as an asymmetric smile but also functional disturbances such as drooling, dribbling, or even dysarthria [12].

It is recommended a total treatment dose of 6 U, which must be divided between two injection sites, one per side. Injections should be performed in the projection of the muscle, 1 cm lateral and 1.5 cm below the oral commissure.

It is very important to avoid injection into neurovascular structures that lie in this area, such as the marginal mandibular nerve and facial artery and vein.

With this aim it could be useful to lift the skin and muscle with the nondominant hand before injecting BOTOX [12].

4.3.3 Mentalis

Its contraction raises the chin, elevates the skin of the lower lip upward, and protrudes and everts the lower lip during drinking.

Its hyperactivity can cause a deep wrinkle between the lower lip and the prominence of the mandible. Moreover, with loss of collagen and subcutaneous fat that occurs with aging, it can appear as chin dimpling.

BOTOX treatment is effective in individuals who exhibit dynamic chin wrinkles [15].

As for the depressor anguli oris, also for mentalis, the dilution should be 100 U vial of BOTOX®

with only 1 ml of normal saline, to avoid the induction of weakness of the depressor labii inferioris, producing in this way muscular aberration of the mouth (inability to speak, drink, and eat) [12].

It is recommended that a total dose of 6 U be divided equally between two injection sites, one per side. Injections should be performed subcutaneously or intramuscularly at two symmetrical points located close to the chin midline, 1 cm above the lower edge of the jaw.

Even if the use of BOTOX®, according to current regulations, is allowed only for glabellar lines, analyzing the existing literature, we can find several scientific articles that show its use for other facial areas, including the perioral one [12–14].

BOTOX injection in the perioral area has two important advantages: it takes only about 20 min to be done; it is a noninvasive technique, with minimal adverse effects if performed by an expert physician.

References

1. Orentreich DS, Orentreich N (1995) Subcutaneous incisionless (subcision) surgery for the correction of depressed scars and wrinkles. Dermatol Surg 21(6):543–549
2. Fernandes D (2005) Minimally invasive percutaneous collagen induction. Oral Maxillofac Surg Clin North Am 17(1):51–63
3. Fabbrocini G, Fardella N, Monfrecola A, Proietti I, Innocenzi D (2009) Acne scarring treatment using skin needling. Clin Exp Dermatol 34(8):874–879
4. Fabbrocini G, De Vita V, Pastore F, Panariello L et al (2011) Combined use of skin needling and platelet-rich plasma in acne scarring treatment. Cosmetic Dermatol 24:177–183
5. McAllister DV, Wang PM, Davis SP, Park JH, Canatella PJ, Allen MG, Prausnitz MR (2003) Microfabricated needles for transdermal delivery of macromolecules and nanoparticles: fabrication methods and transport studies. Proc Natl Acad Sci U S A 100(24):13755–13760
6. Doddaballapur S (2009) Microneedling with dermaroller. J Cutan Aesthet Surg 2:110–111
7. Aust MC, Reimers K, Repenning C, Stahl F, Jahn S, Guggenheim M, Schwaiger N, Gohritz A, Vogt PM (2008) Percutaneous collagen induction: minimally invasive skin rejuvenation without risk of hyperpigmentation-fact or fiction? Plast Reconstr Surg 122(5):1553–1563
8. Fernandes D, Signorini M (2008) Combating photoaging with percutaneous collagen induction. Clin Dermatol 26(2):192–199
9. Meunier FA, Schiavo G, Molgo J (2002) Botulinum neurotoxins: from paralysis to recovery of functional neuromuscular transmission. J Physiol Paris 96(1–2):105–113
10. Nayyar P, Kumar P, Nayyar PV, Singh A (2014) BOTOX: broadening the horizon of dentistry. Clin Diagn Res 8(12):ZE25–ZE29
11. Hsiung GY, Das SK, Ranawaya R et al (2002) Long-term efficacy of botulinum toxin A in treatment of various movement disorders over a 10-year period. Mov Disord 17(6):1288–1293
12. Benedetto AV (2006) Botulin toxin in clinical dermatology. Taylor & Francis
13. Yutskovskaya Y, Gubanova E, Khrustaleva I, Atamanov V et al (2015) IncobotulinumtoxinA in aesthetics: Russian multidisciplinary expert consensus recommendations. Clin, Cosmetic Investigat Dermatol 8:297–306
14. Carruthers JD, Glogau RG, Blitzer A, Facial Aesthetics Consensus Group Faculty (2008) Advances in facial rejuvenation: botulinum toxin type a, hyaluronic acid dermal fillers, and combination therapies – consensus recommendations. Plast Reconstr Surg 121(5 Suppl):5S–30S
15. Beer K, Yohn M, Closter J (2005) A double-blinded, placebo-controlled study of Botox for the treatment of subjects with chin rhytids. J Drugs Dermatol 4(4):417–422

Aesthetic Procedures for Eye Wrinkles: Skin Needling and Botox

5

Gabriella Fabbrocini and Sara Cacciapuoti

5.1 Anatomy of Periocular Region

A good knowledge of periocular region's anatomy is fundamental to aesthetic procedures approach of the eye wrinkles area. This knowledge, including vascular supply, nerve position, and facial compartments, is necessary to provide the best treatment options for patients, manage complications appropriately, achieve optimal results, and avoid unwanted side effects. First of all it is necessary to define anatomic confines of orbit: the superior margin is delimited by the frontal bone and sphenoid; the inferior margin is bordered by the maxilla, palatine, and zygomatic; the medial margin is defined by the ethmoid, lacrimal bone, and frontal bone; and the lateral margin is delimited by the zygomatic and sphenoid. In Table 5.1 we list the bones articulating to form the orbit. Looking more superficially, the skin covering the orbital opening (mostly composed of eyelid skin) is the thinnest in the body, with minimal or no subcutaneous fat. Immediately underlying it is the orbicularis oculi muscle. It is divided into pre-tarsal, pre-septal, and orbital components. Deeper to the orbital part of the orbicularis, along the superior orbital rim, lies the corrugator supercilii muscle. Orbital parts of the orbicularis and corrugator muscle are brow depressors. The orbicularis muscle is the only active force that keeps the lower eyelid margin in its normal position. Deeper to the orbicularis lies the orbital septum, which is a thin fibrous connective tissue layer extending from the orbital rim to the eyelid margin. It is well known that the botulinum target is muscles: in Table 5.2 we summarize the anatomy and function of periocular region muscles [1].

All muscles of the upper face contribute to brow position. This must be considered for the aesthetic appearance of the upper face and must be balanced to achieve an acceptable and pleasing result. All facial muscles of the upper face are innerved by facial nerve (VII cranial nerve). The periocular region is vascularized by branches of the superficial temporal artery, arising from the external carotid artery: the zygomatic-orbital artery (collateral branch of the superficial temporal artery) and the frontal artery (terminal branch of artery temporal surface) [2].

These anatomic basics are essential for a layered approach, a correct evaluation of the skin, fat, muscle, and bone to determine which procedure is best suited for each patient. This chapter evaluates clinical application of two aesthetic procedures available for patients presenting for periorbital rejuvenation: Botox and skin needling.

G. Fabbrocini (✉) • S. Cacciapuoti
Division of Clinical Dermatology, Department of Clinical Medicine and Surgery, University of Naples Federico II, Via Sergio Pansini 5, Naples 80133, Italy
e-mail: gafabbro@unina.it

G. Fabbrocini et al. (eds.), *Nonsurgical Lip and Eye Rejuvenation Techniques*,
DOI 10.1007/978-3-319-23270-6_5

Table 5.1 Bones articulating to form the orbit

Frontal bone (Pars orbitalis)
Lacrimal bone
Ethmoid bone (Lamina papyracea)
Zygomatic bone (Orbital process of the zygomatic bone)
Maxillary bone (Orbital surface of the body of the maxilla)
Palatine bone (Orbital process of palatine bone)
Sphenoid bone (Greater and lesser wings)

5.2 Botulinum and Eye Wrinkles

The use of *Botulinum toxin A* (BoNT-A) in cosmetic dermatology has increased in popularity due to the efficacy and relative safety of the treatment. BoNT-A is a natural substance, made by *Clostridium botulinum* bacteria; it is a very powerful toxin, which causes a temporary paralysis of the muscle. It works on the motor end plates of the skeletal muscle. It also has an action on sympathetic smooth muscle and sweat glands. Several medical companies have developed synthetic forms of the toxin, with the same effects of natural toxin, but safely in microdoses. Some of the mainstream brand names include Botox®, Dysport®, Xeomin®, Vistabel®, and Neurobloc®; these are used to treat periocular wrinkles, brow ptosis, blepharospasm, hemifacial spasm, and facial palsy rehabilitation. The short-term safety profile of BoNT-A in cosmetic nonsurgical procedures was confirmed for all the three commercial formulations. The use of botulinum toxin A (BoNT-A) for aesthetic treatments is growing steadily, and new safety data have been reported in recently published studies [3].

Table 5.2 Anatomy and function of periocular region muscles

Muscles	Anatomy	Functions
Frontalis	This muscle has no bony attachments, as its fibers arise from the scalp occipitofrontalis muscle and aponeurosis and terminate on the skin and dermal tissue of the anterior forehead and brow	The muscle runs in a vertical direction, and in this way contraction will result in horizontal forehead wrinkles above the brow level
Corrugator supercilii	This muscle attaches to the orbital rim medially and inserts with the frontalis on the skin more laterally	Contraction of this muscle produces vertical wrinkles known as "frown lines" in the glabella and lower median forehead
Procerus	The procerus muscle draws down the medial brow by attaching to the facial aponeurosis overlying the nasal bones and inserting on the skin of the eyebrow and lower forehead	Contraction of this muscle produces horizontal wrinkles over the nasal dorsum or "glabellar lines"
Orbicularis oculi orbital part	This is the outermost portion of the muscle that forms a complete ellipse around the bony orbit	This muscle approximates the upper with the lower eyelids, as with forced, volitional eyelid closure, and depresses the medial and lateral aspects of the eyebrow
Orbicularis oculi palpebral part	This muscle is subdivided into pre-septal and pre-tarsal portions. The pre-tarsal portion courses over the eyelids and the pre-septal portion lies superficial to the orbital septum	Contraction of the palpebral part of the orbicularis oculi provides the sphincteric action of the eyelids and gently closes them involuntarily, as occurs with blinking or sleep
Orbicularis oculi lacrimal part	This muscle is located posterior to the medial palpebral ligament and lacrimal sac	Contraction of the lacrimal portion of the orbicularis oculi draws the eyelids posteriorly against the globe, compressing the lacrimal sac and facilitating the lacrimal pump

5.2.1 Mechanism of Action

Since botulism was first described in the eighteenth century, this neurotoxin has undergone a slow development to Botox which is now manufactured. Voluntary muscle contraction is a response to stimulation by action potentials passing along a nerve to the muscle end plate. Once the action potentials reach a synapse at the neuromuscular junction, they stimulate an influx of calcium into the cytoplasm of the nerve ending, and mobilization of acetylcholine toward the synapse occurs. Acetylcholine fuses with the nerve ending membrane and then crosses the synapse to bind with receptors on the muscle fiber, which leads to contraction. BoNT-A inhibits the discharge of acetylcholine into the synapse by bonding to the nerve at the neuromuscular ending. The toxin is then internalized via receptor-mediated endocytosis, and a toxin-containing vesicle is formed within the nerve ending. These internalized vesicles inhibit the acetylcholine protein (synaptosomal-associated protein-25) that is located on the cell membrane. This inhibits muscle contraction, which leads to reversible muscle atrophy (Fig. 5.1).

Physiology and mechanism of action are emphasized because the only way to utilize any type of BoNT-A properly is to have an in-depth understanding of how to modify the normal movements of the mimetic muscles of the face. When injections of BoNT-A are appropriately performed, desirable and reproducible results without adverse sequelae are created. BoNT-A effect starts from between 3 to 7 days after injection and lasts between 2 and 6 months (average 4 months). The peak action is at 7–10 days after injection with complete paralysis of the muscle area treated, which then it gradually wears off. Injections can be repeated [4].

5.2.2 Applications and Technique

As with any other types of treatment, before performing cosmetic procedures with BoNT-A, both the patient and physician should discuss treatment expectations, to prevent disappointment. The areas of dynamic motion, such as the glabella, frontal region, and periorbital lines, are the best application areas for BoNT-A because this procedure is ideal for reduction of mimetic effects of wrinkles and folds, while it is less suitable for static wrinkles and very deep folds.

About the periocular region the best site of application are:

- *Corrugator supercilii muscle,* to correct glabella lines, the vertical frown lines just between and above the two eyebrows. Excellent results are achieved when injecting these lines with BoNT-A. Patients are asked to frown before injection, while the largest muscle body of the procerus and corrugator muscles is palpated. The corrugator supercilii muscle is injected 1 cm above the orbital rim. With a distance less than 1 cm, diffusion of the material into the medial part of the eyebrow is possible and may lead to local eyebrow ptosis. This effect is temporary and not treatable [5]. Female patients require 5 × 4 U (0.1 ml); most males require up to 5 × 6 U (0.15 ml) depending on the muscle tone [6]
- *Orbicularis oculi muscle,* to correct "smile" wrinkles and lines at the outer corners of each eye. Lateral canthal wrinkles are caused by the contraction of the lateral side of the orbital portion of the orbicularis oculi and therefore are referred to as dynamic wrinkles. They are the result of infolding and pleating of the overlying skin, which radiate away from the lateral canthus. These wrinkles are perpendicular to the direction of the lateral muscle fibers of the orbital portion of the orbicularis oculi, which run mostly in a vertical direction around the lateral canthus. Because crow's feet are enhanced during smiling or laughing, the contraction of the risorius and zygomaticus major et minor also contributes to the formation of these lateral canthal wrinkles. Consequently, when persons laugh, smile, or grin, they contract the risorius and zygomaticus major et minor, which also can accentuate the lower aspect of their crow's feet. Three injection sites lateral to each eye are almost always suf-

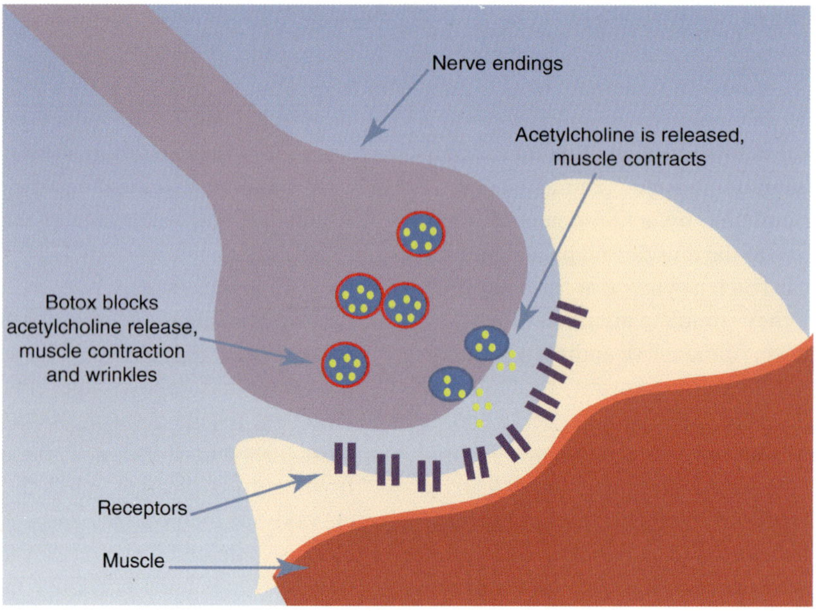

Fig. 5.1 Botulinum mechanism of action

ficient to give relaxation to the part of the orbicularis oculi muscle that is responsible for crow's feet. Each injection site receives 4 U (0.1 ml). This gives adequate response rates up to 16 weeks postinjection [7]. By choosing the upper injection site just below the eyebrow, an aesthetically pleasing lift of the lateral eyebrow can be achieved as a beneficial extra effect of this treatment. The caudal injection site is 1–2 cm below the medial one and stays away from the orbital rim (Fig. 5.2).

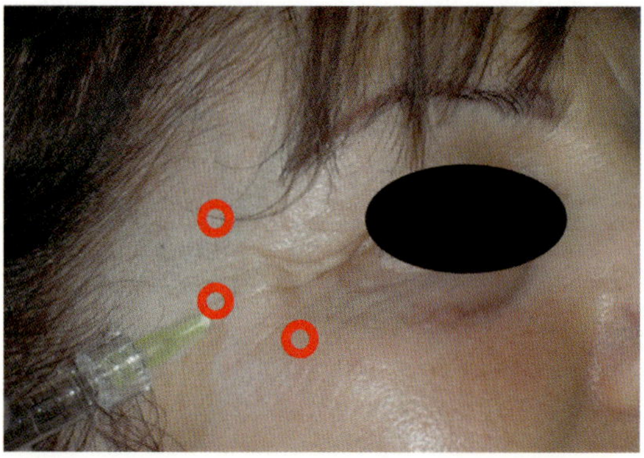

Fig. 5.2 Three injection points are chosen with an equal distribution following the outer rim of the orbit. Distance to lateral orbital rim should be 1 cm

This injection technique gives good aesthetic outcomes with slight elevation of the lateral eyebrow and clear reduction of periorbital lines (Fig. 5.3).

We remember that the palpebral portion of the orbicularis oculi should not be treated with BoNT-A, because it can cause loss of the voluntary and involuntary functions of eyelid closure.

5.2.3 Contraindications and Adverse Effects

Administration of BoNT-A should be avoided during pregnancy and breastfeeding and in patients with disorders of the neuromuscular junction (such as myasthenia gravis, Lambert-Eaton syndrome) and neurodegenerative diseases such as amyotrophic lateral sclerosis. Simultaneous use of aminoglycoside antibiotics (gentamycin, tobramycin) should be avoided because of their potentiating effect on BoNT-A. Other theoretical drug interactions could occur with calcium channel blockers, cyclosporine, and cholinesterase inhibitors. Highly frequent administration of BoNT-A

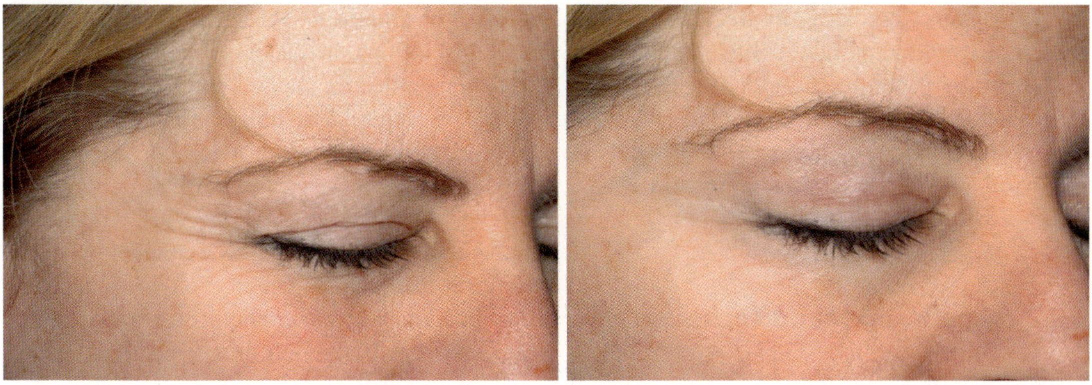

Fig. 5.3 Two weeks before (*left*) and after (*right*) injection with BoNT-A. Slight elevation of the lateral eyebrow and clear reduction of periorbital lines

(more than every 12 weeks) and repeated exposure can lead to formation of neutralizing antibodies against the toxin which lead to disappointing results. Thorough preoperative evaluation with meticulous surgical planning to achieve facial aesthetic balance between the forehead, eyelids, and midface is imperative to avoid or decrease potential functional and/or cosmetic complications in cosmetic periocular surgery. Before performing surgery, the physician should be aware of the patient's history of dry eyes, previous facial trauma, previous injection of Botox Cosmetic, history of previous laser-assisted in situ keratomileusis, and past facial surgery. Intraoperative and postoperative medical and surgical management of cosmetic periocular surgery complications focus on decreasing the risk of postoperative ptosis, lagophthalmos, lid retraction, and lid asymmetry, with special attention to limiting the risk of visual loss secondary to orbital hemorrhage [8].

5.3 Skin Needling and Eye Wrinkles

Skin needling is also called micro-needling therapy or collagen induction therapy. It is a minimally invasive nonsurgical and nonablative procedure for facial rejuvenation that involves the use of a micro-needling device to create controlled skin injury. Skin needling is able to treat wrinkles of the periocular and perilabial region, cheeks, neck, and décolleté. Other areas of the body can also be treated such as back of the hands and arms.

Skin needling can be used for skin rejuvenation: a variety of needle lengths can be used to treat different depths and therefore affect different concerns on the skin. Rollers with longer needles are used on more difficult problems such as deep ingrained wrinkles around the mouth, whereas shorter needles are used for general rejuvenation. In case of wrinkles associated with skin aging, one or two skin needling treatments are recommended every year.

5.3.1 Mechanism of Action

The microneedles penetrate through the epidermis but do not remove it; the epidermis is only punctured and heals rapidly. The needles seem to separate the cells from one another rather than cut through them, and thus many cells are spared. Because the needles are set in a roller, every needle initially penetrates at an angle and then goes deeper as the roller turns. Finally, the needle is extracted at a converse angle, therefore curving the tracts and reflecting the path of the needle as it rolls into and then out of the skin for about 0.5 mm into the dermis. The epidermis, and particularly the stratum corneum, remains intact except for the minute holes, which are about four cells in diameter [9]. The controlled injury triggers the body to fill these microwounds by producing new collagen and elastin in the papillary dermis; in addition, new capillaries are formed. This neovascularization and neocollagenesis following treatment leads to reduction of scars and skin rejuvenation, i.e., improved skin texture, firmness, and hydration.

5.3.2 Applications and Technique

Since 1995, this technique has been used to achieve percutaneous collagen induction in order to reduce skin imperfections [10]. However, to date, skin needling has mostly been proposed as an effective method of treating scars (including hypertrophic scars) caused by acne, surgery or thermal burns, stretch marks, perilabial and periorbital wrinkles, photoaging, and hyperpigmentation, e.g., in melisma [11–14]. For wrinkles of periocular region, skin needling can be considered for the treatment of crown's feet wrinkles and glabella wrinkles with good results. Generally three or four sessions are needed to obtain satisfactory results. Skin needling can be combined at a later stage with other noninvasive procedures such as:

- Lasers
- Photodynamic therapy
- Botox
- Dermal fillers
- Superficial chemical peels

Skin needling is carried out by rolling a special device over the skin comprising a rolling barrel fitted with a variable number of microneedles. There are various skin needling devices including Dermaroller® (Dermaroller GmbH), Dermapen™ (Equipmed Pty Ltd; Australia), Derma-Stamps™ (Dermaroller USA), and radial disks incorporating fine microneedles of various diameter and length, fabricated from a wide range of materials such as silicon, glass, metals, and polymers. The needles are up to 3 mm in length (Fig. 5.4).

The skin needling procedure takes a few minutes up to an hour to complete, depending on the area to be treated and the severity of the problem. No lotions, makeup, or other topical products are applied on the treatment area on the day of the procedure. The skin is punctured in a specific pattern using a skin needling device. The device is rolled over the skin multiple times for best results (Fig. 5.5). As each fine needle punctures the skin, it creates a channel or microwound stimulating skin cell regeneration. Application of local anesthetic cream can prevent procedure pain and help in performing the procedure properly.

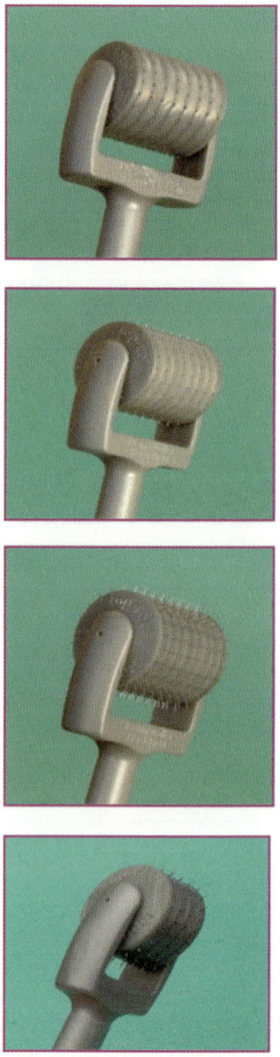

Fig. 5.4 Different skin needling devices with microneedles of various diameter and length

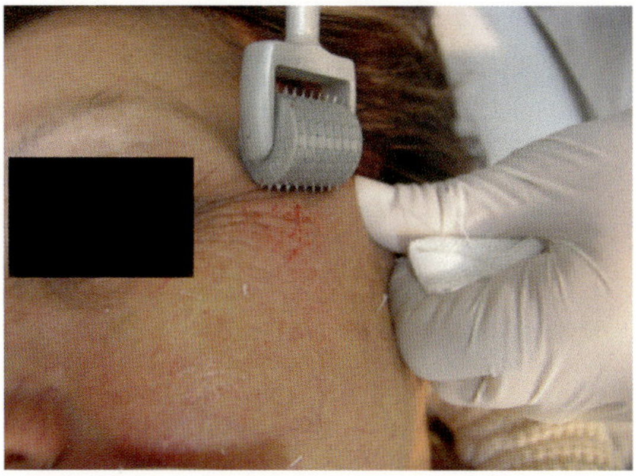

Fig. 5.5 Skin needling procedure for periocular wrinkles: the skin is punctured in multiple directions applying a constant pressure

A minimum of 6 weeks is recommended between two treatments as it takes that long for new natural collagen to form. Skin needling is well tolerated by patients, but dryness, scaling, redness, and swelling may be seen after treatment, lasting for several days or longer, depending on the depth of penetration of the needles. Sun protection for several weeks is recommended. As the microholes close quickly, postoperative wound infection is rare. Emollients or antibiotic creams may be prescribed, if considered necessary. Rejuvenation of skin may be seen as soon as 2 weeks and as long as 6–8 months after the medical procedure. The number of needling procedures depends on the individual skin condition. Three to four treatments may be needed for moderate acne scars. (Fig. 5.6)

5.3.3 Contraindications and Adverse Effects

Some clinical condition can be considered as absolute contraindication to skin needling procedure:

- Treatment with Roaccutane within the last 3 months
- Presence of open wounds, cuts, or abrasions on the skin
- Radiation treatment within the last year
- A current outbreak of herpes simplex or any other infection or chronic skin condition in the area to be treated
- Areas of the skin that are numb or lack sensation
- Pregnancy or breastfeeding
- History of keloid or hypertrophic scars or poor wound healing

The observation of all the pre- and postoperative precautions and respect of contraindication reduce the risk of adverse effects that are minimal with this type of treatment and typically include minor flaking or dryness of the skin, with scab formation in rare cases, milia, and hyperpigmentation which can occur only very rarely and usually resolves after a

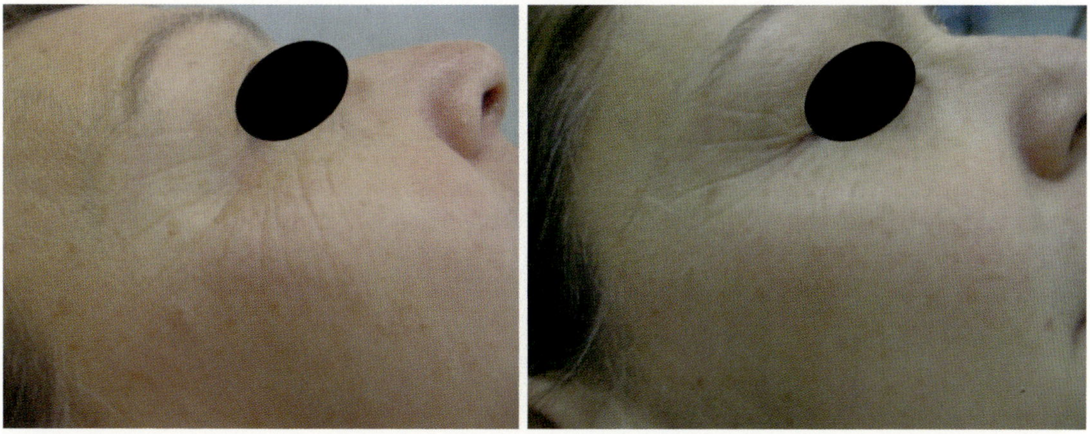

Fig. 5.6 Periocular wrinkles in a 52 years old patient before (*left*) and after (*right*) four skin needling sessions

month. Edema and erythema are the most frequent sequelae. Recovery may take 24 h or up to a few days. Most patients are able to return to work the following day. Recovery time depends on the treatment level and the length of the needles.

Conclusions

Botox and skin needling are two procedures that can give good aesthetic outcome in periorbital wrinkles correction. Professional skin needling is considered to be one of the safest skin treatment procedures for this area. Unlike chemical peels, dermabrasion, and laser treatments, skin needling causes minimal damage to the thin skin of the periocular region. Botox has an ideal characteristic: it is the best treatment to correct dynamic wrinkles as periocular ones. Moreover if a complication does arise, while not aesthetically acceptable and potentially untoward, it is time limited, and the anatomical area will eventually return to its pretreatment baseline status. For these reasons, Botox and skin needling can be considered two useful tools that physicians can use for periocular wrinkle correction.

References

1. Shams PN, Ortiz-Pérez S, Joshi N (2013) Clinical anatomy of the periocular region. Facial Plast Surg 29(4):255–263
2. Love LP, Farrior EH (2010) Periocular anatomy and aging. Facial Plast Surg Clin North Am 18(3): 411–417
3. Cavallini M, Cirillo P, Fundarò SP et al (2014) Safety of botulinum toxin A in aesthetic treatments: a systematic review of clinical studies. Dermatol Surg 40(5):525–536
4. Klein AW (1998) Dilution and storage of botulinum toxin. Dermatol Surg 24:1179–1180
5. Carruthers A, Carruthers J (2007) Eyebrow height after botulinum toxin type A to the glabella. Dermatol Surg 3:S26–S31
6. Carruthers A, Carruthers J (2005) Prospective, double-blind, randomized, parallel group, dose-ranging study of botulinum toxin type A in men with glabellar rhytids. Dermatol Surg 31:1297–1303
7. Lowe NJ, Ascher B, Heckmann M et al (2005) Double-blind, randomized, placebo-controlled, dose-response study of the safety and efficacy of botulinum toxin type A in subjects with crow's feet. Dermatol Surg 31:257–262
8. Pena MA, Alam M, Yoo SS (2009) Complications in fillers and Botox. Oral Maxillofac Surg Clin North Am 21(1):13–21
9. McAllister DV, Wang PM, Davis SP et al (2003) Microfabricated needles for transdermal delivery of macromolecules and nanoparticles: fabrication methods and transport studies. Proc Natl Acad Sci U S A 100(2):13755–13760
10. Orentreich DS, Orentreich N (1995) Subcutaneous incisionless (subcision) surgery for the correction of depressed scars and wrinkles. Dermatol Surg 21(6):543–549
11. Fabbrocini G, Fardella N, Monfrecola A et al (2009) Acne scarring treatment using skin needling. Clin Exp Dermatol 34(8):874–879
12. Fabbrocini G, De Padova MP, De Vita V et al (2009) Trattamento de ruga periorbitais por terapia de inducao de colageno. Surg Cosmetic Dermatol 1(3):106–111
13. Fabbrocini G, De Vita V, Fardella N et al (2011) Skin needling to enhance depigmenting serum penetration in the treatment of melasma. Plast Surg Int 2011:158241
14. Fernandes D (2005) Minimally invasive percutaneous collagen induction. Oral Maxillofac Surg Clin North Am 17(1):51–63

Chemical Peeling for the Lip and the Eye Regions

6

Aurora Tedeschi

6.1 Introduction

Chemical peeling, or chemoexfoliation, is a dermatological procedure commonly used for both skin rejuvenation and some cutaneous conditions. It consists of the application of one or more chemical exfoliating compounds to the skin to remove and regenerate part of the epidermis and dermis. This may result in the improvement of the physical appearance of the skin and a decrease in the number of wrinkles, pigmentations (e.g., melasma, lentigo), and inflammatory lesions (e.g., acne, rosacea) [1, 2].

Many chemical compounds may be used as peeling agents and their effects may differ, varying from light to medium and deep regeneration. This review focuses on the use of peeling in the perioral and periocular region for the treatment of aging, photoaging, and melasma.

6.2 History

Although chemoexfoliation represents one of the oldest cosmetic procedures, described in ancient Egyptian writings [3] and performed by Romans as well as Indian and Turkish women through rudimentary methods to smooth the skin, the concept of peeling acquired a scientific identification in the late 1800s, when phenol was described to lighten the skin [4]. Soon other peeling agents (e.g., salicylic acid, resorcinol, trichloroacetic acid) were identified, and in the mid-1900s peeling procedures were used for medical purposes. However, it was only in the 1970s that chemical peeling became popular [1, 5].

6.3 Classification

Before focusing on specific peeling for the lip and eye region, a brief classification of generic chemical peeling agents is listed based on of the depth of dermal penetration [1, 5]:

1. Very superficial (*glycolic acid 30–50 %, Jessner solution applied in 1–3 coats, salicylic acid 25 % applied in 1 coat, resorcinol, 20 % applied briefly (5–10 min), trichloroacetic acid (TCA), 10 % applied in 1 coat*), in which the area to be treated is confined to the stratum corneum, with no alteration below it
2. Superficial (*glycolic acid, 50–70 % applied for 3–10 min; salicylic acid 25 % applied in 4–10 coats, pyruvic acid 40 % applied in 4–5 coats, Jessner solution applied in 4–10 coats; resorcinol, 40 % applied for 30–60 min; TCA 20 %*), when part or all of the epidermis is involved

A. Tedeschi
Department of Dermatology, Dermatology Clinic, University of Catania, Catania, Italy
e-mail: auroratedeschi@gmail.com

G. Fabbrocini et al. (eds.), *Nonsurgical Lip and Eye Rejuvenation Techniques*,
DOI 10.1007/978-3-319-23270-6_6

3. Medium (*TCA 35 %, pyruvic acid 50–60 % applied in several coats, augmented TCA (glycolic acid 70 % + TCA 35 %, Jessner solution + TCA 35 %, salicylic acid + TCA 35 %)*, involving both epidermidis and papillary dermis
4. Deep (*TCA 50 %, phenol*), involving the epidermis, papillary dermis, and reticular dermis

Very superficial and *superficial peelings* involve the stratum corneum or the epidermis in toto and represent well-tolerated treatments with very low risk of side effects.

Medium-depth peelings involve epidermis and papillary dermis, causing denaturalization of proteins, clinically characterized by skin bleaching (frosting). Histologic modifications in connective tissue with new deposition of collagen and elastic fibers may be observed after these procedures. They require a posttreatment procedure and are associated with some side effects.

Deep peelings cause a significant dermal injury, involving the reticular layer. They also cause a quick and intense frost, resulting in dermal regeneration with new deposition of collagen and glycosaminoglycans. Special care is required for this type of peeling since severe complications may occur.

The choice of the most appropriate peeling agent, keeping in mind the related depth of penetration in the skin, is crucial.

Both perioral and periocular regions represent very sensitive anatomic areas and, therefore, in general the use of soft peelings is recommended. This chapter reviews the types of peeling indicated for these specific areas.

6.4 Considerations

Before considering chemoexfoliation, a series of evaluations should be made. First of all, a thorough patient evaluation including age, sex, skin type, aging, and photoaging severity, in addition to the presence of any psychological discomfort or other skin disorders, must be considered. Moreover, the patient's history of abnormal or keloid scarring, perioral herpes simplex virus (HSV) infection, prior treatments such as oral isotretinoin, radiation, or laser skin resurfacing, and photosensitizing medications should also be carefully evaluated to avoid scarring or slow reepithelialization [1, 5].

Skin type and phototype should be also carefully examined. Thicker and oily skins, for instance, are more resistant to peeling and may require a deeper treatment than other skin types. Fitzpatrick's phototypes IV–VI are not recommended for medium to deep peeling because of the high risk for pigmentary dyschromias. A positive history for other skin disorders, such as atopic dermatitis, seborrheic dermatitis, psoriasis, contact dermatitis, or rosacea must be investigated for their potential exacerbation during the postpeeling period. Patients with a history of HSV should be treated with antiviral drugs from the prepeel period until complete reepithelialization, especially when medium-depth or deep peelings are performed [5]. Patients with significant history or current evidence of any psychological disorder, or with immunocompromising diseases or allergies, should not be treated.

Skin priming with topical compounds (retinoic acid, glycolic acid, pyruvic acid, and hydroquinone) is usually suggested 2 weeks before the peeling to improve its performance. Skin priming allows an easier and uniform penetration of the peeling agent, reducing the reepithelialization phase as well as the risk of posttreatment hyperpigmentation. However, when treating the periocular and/or perioral area, skin priming should be avoided as it may irritate the skin.

Finally, complications of peeling should be considered before performing a peel, keeping in mind the direct relationship between the frequency of complications and the peel's depth (deeper treatments lead to more complications) [5–7]. The most frequent changes are pigmentary (hyperpigmentation and hypopigmentation). Phototypes IV–V are at higher risk especially when medium-depth peeling is performed. Early sun exposure and/or the use of oral contraceptives are aggravating factors [5, 8, 9]. Scarring (atrophic or hypertrophic scars) represents a relevant complication for deep peelings. Scars usually appear on the lower part of the face (perioral region), probably due to the mechanical stretching occurring in this area during eating and speaking. Lower eyelid ectropion has also been observed 3–6 months after phenol peeling [5, 9].

HSV represents a frequent complication in patients with a history of HSV recurrence when undergoing a medium-depth peeling. HSV prophylaxis is necessary when these procedures are performed but not for superficial peelings [5]. Bacterial infections are not common but can be observed, *Pseudomonas* infection is the most problematic. Other possible pathogens include *Staphylococcus*, *Streptococcus*, and *Candida* [5].

Persistent erythema is considered a physiological event when skin remains erythematous for up to 3 weeks after the peel. Erythema is caused by angiogenic factors stimulating vasodilatation [3]. When erythema persists for more than 3 weeks and is associated with pruritus, it could be indicative of scarring formation, requiring the use for a short period of potent topical corticosteroids or systemic steroids. Silicone sheeting or pulsating dye laser represent other therapeutic options, especially in cases of evident thickening or scarring [1, 3].

Milia may occur after a period of 8–16 weeks after a procedure, probably resulting from occlusive postpeeling treatments.

Acneiform eruption may be observed in a small percentage of patients during the reepithelialization phase or immediately after, owing to an exacerbation of preexisting acne-prone skin or the use of occlusive products on the skin during the postpeeling period [1]. Systemic antibiotics are usually administered to obtain satisfactory results.

Allergic reactions are relatively rare and most commonly associated with the use of resorcinol. Allergic reactions may be misdiagnosed as the clinical presentation (erythema, pruritus, edema) resembles normal postpeeling reactions. Antihistamines together with steroids may be used to manage these complications.

Cardiotoxicity is a potentially severe complication that may occur during phenol peeling. It has been demonstrated that phenol can be responsible for cardiac toxicity, including tachycardia (arrhythmia), premature ventricular beats, bigeminy, and atrial and ventricular tachycardia [7, 10], in addition to liver and kidney side effects [9]. It is therefore important that a phenol peel is performed by qualified physicians in an operating room with cardiopulmonary monitoring of the patient [7].

6.4.1 Glycolic Acid

Alpha-hydroxy acids (AHA) are a group of carboxylic acids characterized by a hydroxyl group attached to the alpha position of the carbon atom. AHAs increase epidermal thickness and dermal glycosaminoglycan content and are used to treat photoaging, acne, pigmentary, and keratinization disorders [11]. Glycolic acid represents one of the most commonly used AHAs, used both in topical creams at low concentrations (5–20 %) and as a peeling agent at concentrations up to 70 %. Because of its small molecular weight and size, it has high skin penetration. It is considered a relatively safe, effective, and well tolerated peeling agent. Glycolic acid causes superficial peeling with few complications, although dermal wounds similar to those caused by 40 % TCA have been reported with the use of higher concentration (70 %) or in cases of prolonged exposure [8]. Neutralization with any alkaline solution (generally sodium bicarbonate 8–15 %) is required after glycolic acid peeling to avoid any further penetration through the deeper skin layers. Glycolic acid can be used for both perioral and periocular regions, avoiding deep penetration or long exposure and providing prompt neutralization with a solution of sodium bicarbonate as soon as erythema appears. Particular care should be taken for vulnerable areas, such as nasal ala, lips, lateral canthus, and oral commissures, which can be protected with an ointment. Treatment can be repeated every 3–4 weeks for a total of six treatments.

Later on, moisturizers, emollients, and sunscreens must be applied for 5–7 days during the healing process. No other particular medication is required; in addition, cream containing AHAs should be avoided for 2–3 days following the procedure [1, 5].

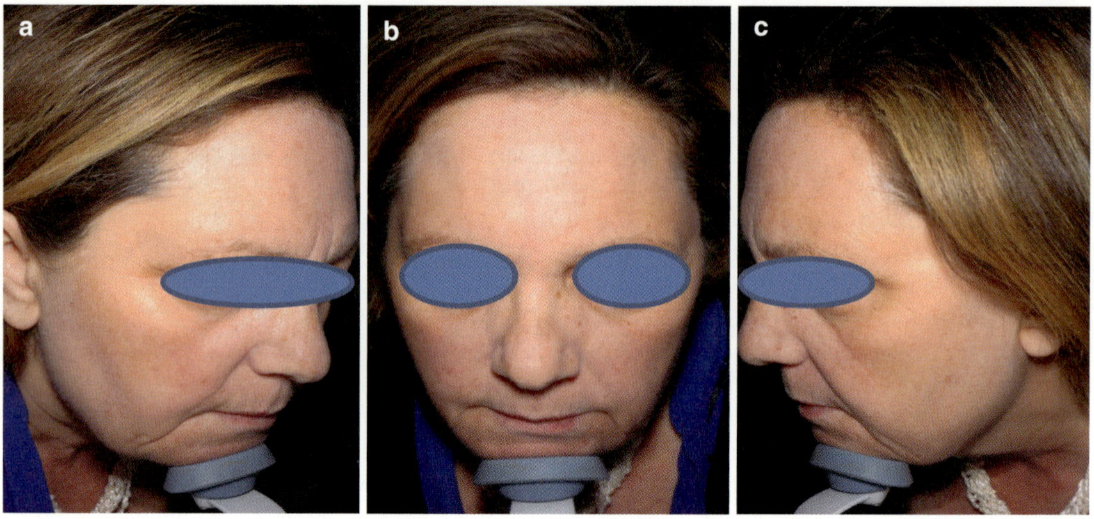

Fig. 6.1 Before salicylic acid peel 30 %

6.4.2 Mandelic Acid

Mandelic acid is an AHA derived from almonds. The large size of the molecule causes a slow and uniform penetration, making mandelic acid, at 30–50 %, the ideal treatment for sensitive skins. Although it is not specific for perioral and periocular areas, it can be used with no precautions or restriction in these areas [5].

6.4.3 Tretinoin Peel

Tretinoin peel is a solution of tretinoin at a high concentration, varying from 1 to 5 %, in propylene glycol. It is applied by gauze or a brush in one or more coats and left on the skin for 4–8 h, after which it is removed with water [6, 12, 13]. In consideration of its potential irritative effects, this kind of peeling should be managed with particular care in the perioral area affected by melasma, restricting the time of exposition. Because of its teratogenicity, this type of peeling should be avoided in women at any stage of pregnancy.

6.4.4 Salicylic Acid

Salicylic acid is an organic carboxylic acid with a hydroxyl group in the beta position [6, 14]. Its lipophilic structure allows it to easily penetrate through sebaceous glands and corneous cells, with consequent destruction and exfoliation of the upper layers of the epidermis [5]. It is mainly indicated for superficial and medium acne scars, particularly for those with a remarkable hyperchromic component; inflammatory acne, rosacea; melasma; and photoaging [14] (Figs. 6.1 and 6.2). It is not a specifically designed peel for perioral and periocular areas, but can be used in these regions with some precautions. Patients suffering from salicylate allergy cannot receive salicylic acid as a peeling treatment [5].

6.4.5 Yellow Peel

Yellow peel (YP) is a combination of retinoic acid with phytic acid, kojic acid, and azelaic acid, which block the synthesis of melanin at different levels. Vitamin C, bisabolol, and salicylic acid are also contained in this formulation. The name of this peel originates from the particular yellow coloration of the skin after its application. YP allows a sort of modulating peeling involving both superficial and medium epidermis. Furthermore, it induces new epidermis regeneration with few risks for potential dyschromia. Mainly indicated for melasma and hyperpigmentation in general [5, 15], it can be used in perioral and periocular areas, using a gentle massage and leaving on the skin no longer than 15–30 min.

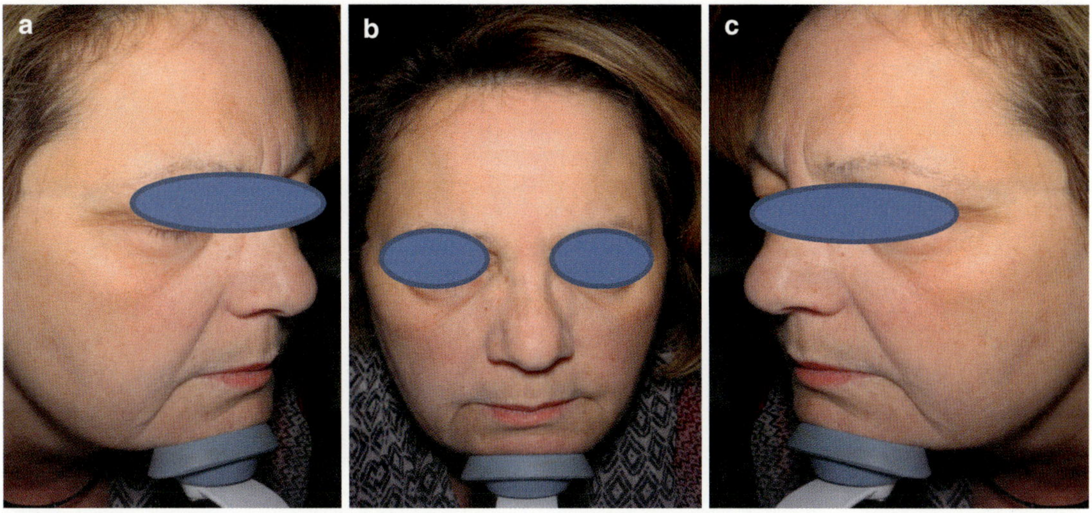

Fig. 6.2 Five days after salicylic acid peel 30 %

6.4.6 Resorcinol Peel

Resorcinol, or m-hydroxybenzene, is a compound structurally and chemically similar to phenol [13]. It is a reducing agent, used in concentrations ranging from 10 to 50 % [1], and is able to break keratin bonds and induce both epidermal regeneration and dermal fibroblast proliferation. Resorcinol is usually applied with a spatula and left on the skin for 25–60 min, increasing by 5 min each week, according to different authors [5, 7, 16]. After the paste has been removed, the skin appears burned and exfoliates for the following 7–10 days. Postpeel care with antibiotic and corticosteroid creams together with sunscreen is important to prevent complications (pigmentary changes and allergic reactions). The main indications for a resorcinol peel are acne, including comedonic acne, along with pigmented lesions, melasma, and superficial scars [5]. It can be used in the perioral area affected by melasma.

6.4.7 Jessner's Solution

Jessner's solution (JS) is a combination of salicylic acid (14 g), resorcinol (14 g), lactic acid (85 %, 14 g) and ethanol (95 %, up to 100 ml), which can be used either alone for superficial peeling or in combination with other agents to facilitate medium-depth procedures [5, 7, 17]. Its efficacy in the treatment of comedonic and inflammatory acne and dyschromias depends on both its keratolytic and anti-inflammatory activity [8, 17]. JS causes keratinocyte dyscohesion as well as intra- and intercellular edema. JS is usually applied in two to three coats with wet gauze, sponges, or a brush. The application of the solution is typically accompanied by mild erythema and an intense burning sensation, followed by a faint frost presenting as a whitening of the skin with a dust-like aspect. Postpeel exfoliation usually occurs within few days and may persist for up to 8–10 days [5]. It can be used in the perioral area affected by melasma.

6.4.8 Trichloroacetic Acid

Among medium-deep peeling, trichloroacetic acid (TCA) represents a good option for perioral or periocular areas affected by advanced photoaging. Usually used at concentrations ranging from 10–20 % to 35–50 % for superficial and medium-depth peeling, respectively [5, 8], the use of TCA in concentrations higher than 35 % is not suggested because of potential scarring. The procedure may be painful. TCA can be applied with cotton tips, brushes, or small gauzes. Peeling depth is easily monitored by erythema and frosting degrees. In particular, minimal erythema represents a very superficial peeling, involving

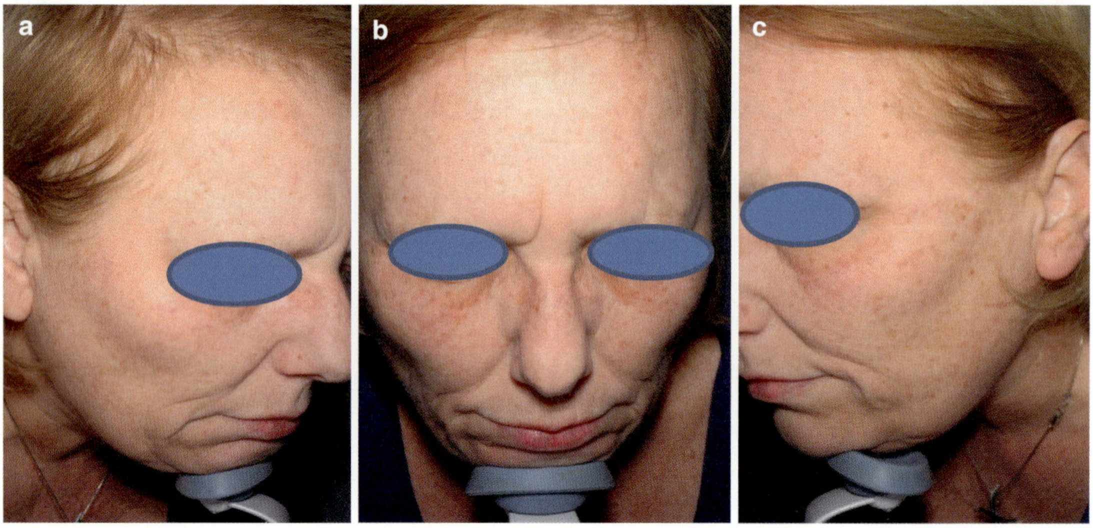

Fig. 6.3 Before TCA peel 25 %

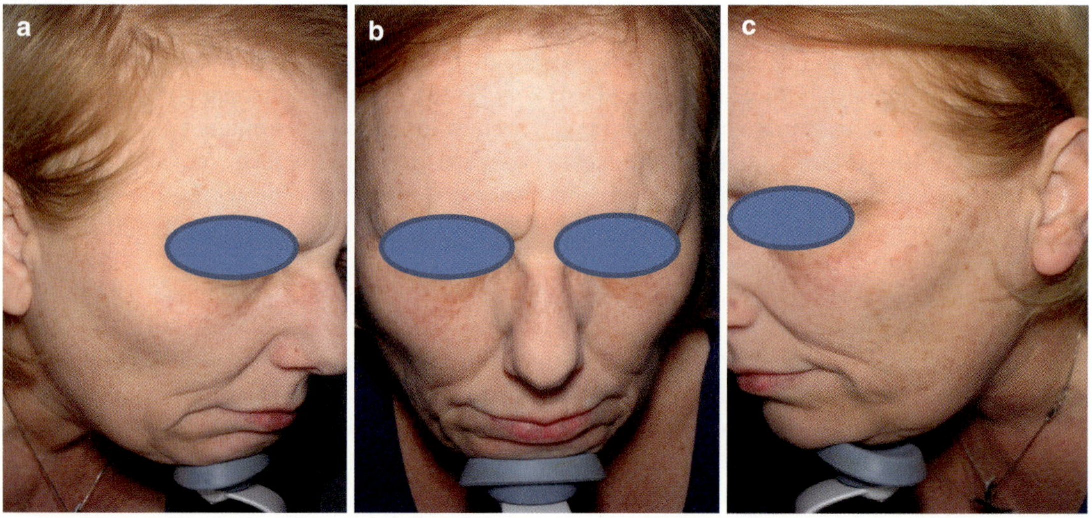

Fig. 6.4 Five days after TCA peel 25 %

mostly the stratum corneum; mild erythema with light frosting patches corresponds to superficial peeling, causing 2–4 days of exfoliation. White frost with a background of erythema shows a medium-depth peel and solid white frost is indicative of a deep peel, extending down the papillary dermis [5, 7, 8]. When TCA is applied in several coats, a deeper peeling is obtained. In such instances it is better to use lighter TCA concentrations. An intense burning sensation is typical of TCA peelings and requires the use of wet cold compresses at the end of the procedure. Afterward, a cream or ointment with 1 % hydrocortisone may be applied to soothe the skin. Sun exposure must be avoided for 4–5 months after the procedure. Patients should be informed about darkening skin color and potential swelling. The exfoliation usually begins 3–4 days after the peeling procedure. During this period the skin must not be removed to avoid postinflammatory hyperpigmentation. If erythema persists for 2 or 3 weeks after exfoliation, the use of light corticosteroid or zinc oxide paste is suggested (Figs. 6.3 and 6.4).

TCA is variably responsible for changes in epidermal thickness, epidermal and dermal protein denaturation, and coagulative necrosis, resulting in epidermis revitalization, an increase in both fibroblasts and collagen types I and III, and reduction of the elastic component [7]. TCA should be avoided in patients with phototype V–VI because of potential darkening and scarring [18].

An innovative formulation of TCA, 3.75 % combined with lactic acid 15 %, specifically designed for periocular and perioral areas, has recently become available. This combination peel was successfully used to improve periorbital hyperpigmentation with very low risk [19].

Conclusion

Perioral and periocular peelings, in consideration of the anatomic areas characterized by thin skin and sensitive skin, should be soft and never aggressive.

References

1. Tedeschi A, Massimino D, Fabbrocini G, Micali G (2012) Chemical peelings. In: Scuderi N, Toth BA (eds) Plastic surgery. Springer, Berlin Heildeberg
2. Tosti A, De Padova MP, Verzì AE, Tedeschi A (2013) Chemical peelings. In: Schwartz RA, Micali G (eds) Acne. Macmillan, Medical Communications, Gurgaon
3. Brody HJ, Monheit GD, Resnik SS, Alt TH (2000) A history of chemical peeling. Dermatol Surg 26:405–409
4. Berardesca E, Cameli N, Primavera G, Carrera M (2006) Clinical and instrumental evaluation of skin improvement after treatment with a new 50% pyruvic acid peel. Dermatol Surg 32:526–531
5. Tedeschi A, Massimino D, Fabbrocini G, West L, De Padova MP, Micali G (2003) Chemical peelings. In: Scuderi N, Toth B (eds) International textbook of aesthetic surgery. Verduci Editore, Roma, in press
6. Landau M (2008) Chemical peels. Clin Dermatol 26:200–208
7. Ghersetich I, Brazzini B, Lotti T, De Padova MP, Tosti A (2006) Resorcinol. In: Tosti A, Grimes PE, De Padova MP (eds) Color atlas of chemical peels. Springer, Berlin Heidelberg, pp 41–47
8. Clark E, Scerri L (2008) Superficial and medium-depth chemical peels. Clin Dermatol 26:209–218
9. Tedeschi A, Massimino D, West L, Micali G (2012) Management of the patients. In: Tosti A, Grimes PE, De Padova MP (eds) Atlas of chemical peels. Springer, Berlin Heidelberg
10. Park JH, Choi YD, Kim SW, Kim YC, Park SW (2007) Effectiveness of modified phenol peel (Exoderm) on facial wrinkles, acne scars and other skin problems of Asian patients. J Dermatol 34:17–24
11. Fabbrocini G, De Padova MP, Tosti A (2006) Glycolic acid. In: Tosti A, Grimes PE, De Padova MP (eds) Color atlas of chemical peels. Springer, Berlin Heidelberg, pp 13–21
12. Cucé LC, Bertino MC, Scattone L, Birkenhauer MC (2001) Tretinoin peeling. Dermatol Surg 27:12–14
13. Khunger N, Task Force IADVL (2008) Standard guidelines of care for chemical peels. Indian J Dermatol Venereol Leprol 74:5–12
14. Grimes PE (2006) Salicylic acid. In: Tosti A, Grimes PE, De Padova MP (eds) Color atlas of chemical peels. Springer, Berlin Heidelberg, pp 49–57
15. Gupta AK, Gover MD, Nouri K, Taylor S (2006) The treatment of melasma: a review of clinical trials. J Am Acad Dermatol 55:1048–1065
16. Ghersetich I, Teofoli P, Gantcheva M, Ribuffo M, Puddu P (1997) Chemical peeling: how, when, why? J Eur Acad Dermatol Venereol 8:1–11
17. Grimes PE (2006) Jessner's solution. In: Tosti A, Grimes PE, De Padova MP (eds) Color atlas of chemical peels. Springer, Berlin Heidelberg, pp 23–29
18. Camacho FM (2005) Medium-depth and deep chemical peels. J Cosmet Dermatol 4:117–128
19. Vavouli C, Katsambas A, Gregoriou S, Teodor A, Salavastru C, Alexandru A, Kontochristopoulos G (2013) Chemical peeling with trichloroacetic acid and lactic acid for infraorbital dark circles. J Cosmet Dermatol 12(3):204–209

Radiofrequency Therapy

7

Patrizia Forgione

7.1 Introduction

Radiofrequency (RF) as used in the field of medicine refers to the use of electromagnetic waves at radio frequencies to produce heat.

RF has been used in surgical and non-surgical aesthetic medicine for more than 70 years. Radio frequencies between 30 and 30 MHz are generally used to produce heat at various skin levels, and hence can be used in the treatment of skin laxity and cellulite [1].

RF treatment makes use of a fundamental concept in the theory of electricity, namely, if a given quantity of electric current encounters a resistance (for DC current) and/or impedance (for AC current), heat is produced in direct proportion to the current and the resistance and/or impedance.

If RF treatment is used to destroy tissue it is referred to as "ablative RF." If it does not destroy tissue it is referred to as "non-ablative RF".

7.2 Ablative RF

RF ablation procedures most commonly employ a radio scalpel when it is necessary to destroy a limited volume of skin in a controlled and reproducible manner.

P. Forgione, MD
Department of Dermatology, Ascalesi Hospital, Naples, Italy
e-mail: forgionepatrizia@virgilio.it

RF ablation therapies are used to treat neoplasms and hepatic, pancreatic, bone, and pulmonary metastasis by means of thermoablation. They have also been successfully used in treating pain and cardiac arrhythmia.

It is noteworthy that the introduction of an RF-ablative probe directly into a blood vessel permits the closure and destruction of the target vessel in much the same way as does laser therapy. This technique is recommended for treating rectilinear veins such as the great saphenous vein.

7.3 Non-ablative RF

Non-ablative RF procedures have many fields of application in aesthetic medicine, especially where the common denominator is the treatment of skin laxity.

In November 2001, the Food and Drug Administration (FDA) approved RF therapy as the preferred method for the treatment for skin depressions [1].

The so-called "aesthetic" effects of RF therapy operate on thermally damaged collagen and elastin by breaking their intermolecular bonds, which in turn brings about a re-structuring of the dermal fibers over the following weeks and, hence, elasticizing of the skin.

G. Fabbrocini et al. (eds.), *Nonsurgical Lip and Eye Rejuvenation Techniques*,
DOI 10.1007/978-3-319-23270-6_7

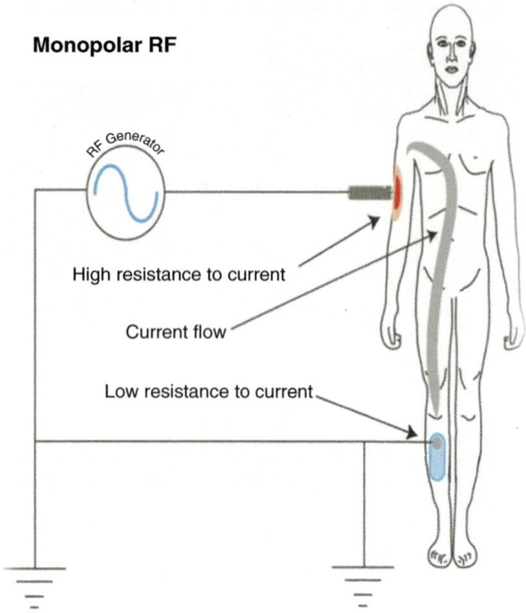

Fig. 7.1 Monopolar RF

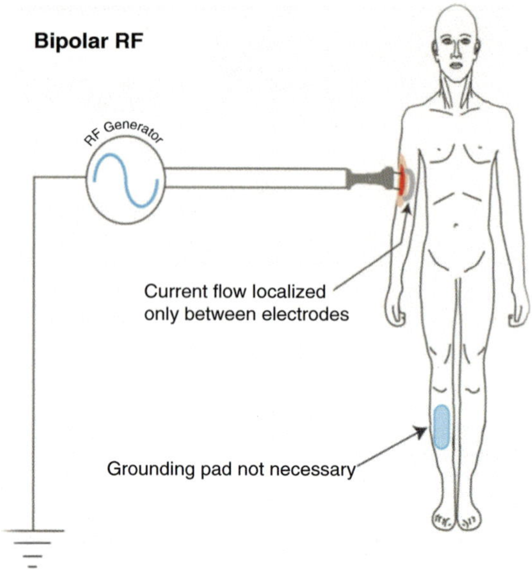

Fig. 7.2 Bipolar RF

7.4 Non-ablative Instrumentation

Monopolar RF handsets apply RF energy generated by an RF unit to one point of the skin to be treated, and a metal plate, also connected to the RF generator, is placed at another point.

Until recently the handsets used in non-ablative therapies have been of two types, monopolar and bipolar, but tripolar and quadripolar handsets are now available for the application of RF therapy [2, 3].

The RF monopolar handset requires a metal plate to be placed near the area to be treated by the handset (Fig. 7.1).

The RF bipolar handset does not require the use of a metal plate, since both poles of the RF generator pulses are in the handset itself (Fig. 7.2).

7.5 General Effects of RF on Skin

The impedance that an RF circuit encounters depends of the conductivity of the tissue to be treated, e.g., the thickness of dermis, the quantity of fat, and the thickness and structure of the connective and glandular tissue.

The higher the impedance of the tissue to be treated, the greater are the heat produced and the thermal effects [4].

Furthermore, owing to the impedance of the skin, the deeper the penetration of the RF electromagnetic wave, the more heat it produces. The heat induced by the RF wave travels from the skin surface to the hypodermis producing temperatures that vary from 30 to 35 °C at the surface, and from 60 to 65 °C at 9 mm.

The increase in temperature causes an increase in blood flow through vessels, leading to drainage in the adipose tissue. In addition, the heat generated by this increase in temperature shrinks the collagen fibers, creating a progressive regenerating effect during the following weeks.

New collagen results from the heat produced by RF waves through a series of intermediary steps. The RF heat produces mediating heat-shock proteins, which in turn stimulate the T lymphocytes and monocytes to produce cytokines and fibroblast growth factor 1, which in turn stimulates the fibroblasts to produce the new collagen.

Histological studies conducted 4 months after treatment showed an increase in collagen density and a reduction in volume of the sebaceous glands.

7.6 Indications

Numerous studies have shown that RF treatment can be safely used on both dark and light phototypes since it does not interfere with melanogenesis.

RF treatment is indicated for skin laxity of the face and body, and has recently been indicated for the treatment of cellulite and stretch marks [4, 5].

In addition, more recent studies indicate that RF can reduce hyperhidrosis.

Contraindications are pregnancy, arrhythmia, use of a pacemaker, epilepsy, anticoagulant drug use, and skin infections.

7.7 Methodology: Post-treatment Effects and Complications of RF

The treatment protocol prescribes one or more passes over the area to be treated depending on the complexity of the case, and each therapy session can last from 15 to 35/45 min.

The treatment requires the application of a simple gel, or a gel to which either hyaluronic acid or other active ingredients have been added to enhance the anti-aging effect (Fig. 7.3).

The heat transmitted by RF therapy can provoke an unpleasant sensation depending on the intensity of the heat, the area treated, and the sensitivity of the patient.

Immediately after treatment a slight erythema may appear over the treated area (in 35 % of patients), which disappears after 2–3 h and allows the patient to continue with normal daily activity (Fig. 7.4).

It should be noted that there are now systems for the cooling of skin that limit the dispersion of heat toward the epidermis and hence reduce the risk of skin burn.

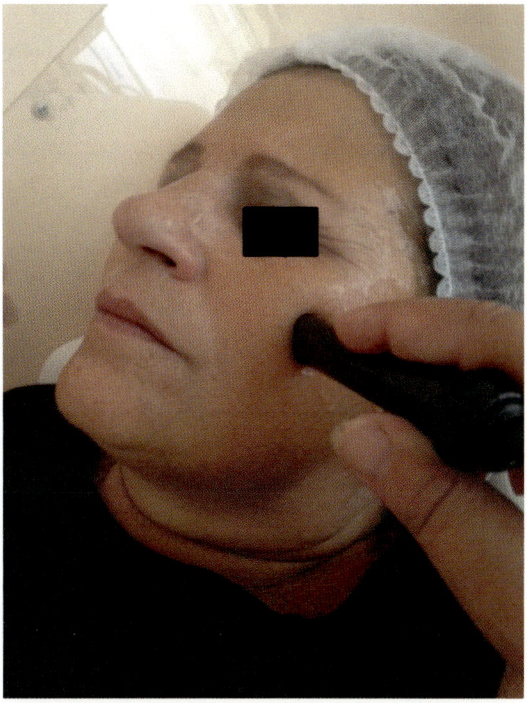

Fig. 7.3 RF handset (nonfractional)

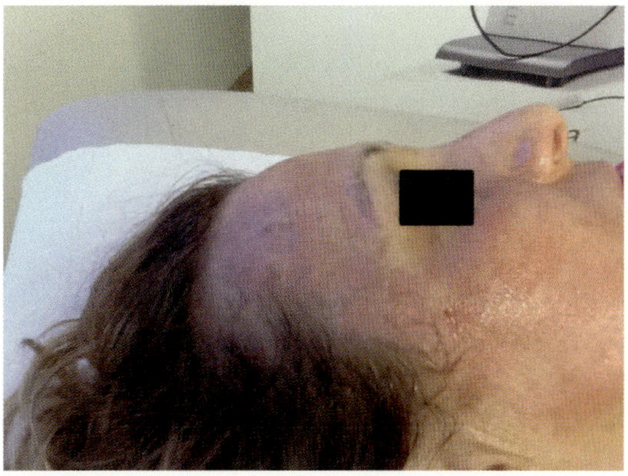

Fig. 7.4 Postinflammatory erythema treatment

It is advisable to apply a lenitive mask after treatment and subsequently a repair cream together with a strong sunscreen.

Very rarely tiny vesicles form, which subsequently transform into crusts that exfoliate in 4–5 days. To facilitate this process one can apply a hyaluronic acid-based cream. It is important to note that the patient should not be exposed to the sun in the 2 weeks following treatment, and in any case should apply a strong sunscreen whenever outdoors.

7.8 Advanced Applications: Fractional RF

Fractional RF, termed "fractional" in analogy with the fractional laser, produces short, intense electric pulses at adjustable frequencies that travel from one electrode to another on the handset, thus generating heat on/in the skin to be treated [6, 7].

Bipolar RF is one of the most innovative, noninvasive treatments for the face and body in the field of aesthetic dermatology [8].

Skin applications of fractional RF have two principal effects:

1. Selective vaporization of superficial layers of skin less than 1 mm
2. A series of microholes in the skin which, upon healing, produce tighter skin.

The distance between holes can be varied by varying the frequency of impulse repetitions; the size and depth of the holes can be varied by varying the energy generated and emitted by the handset.

7.9 Handset Characteristics and Effects on the Skin

There are many types of handset used in fractional RF, which can have from a minimum of 5 to a maximum of 225 microneedles. Treatment with these microneedles causes a series of damaged micropoints on the skin, which initiate a regenerative process.

The needles are equipped with "shock absorbers," which allow the technician to adapt to all facial and body contours while maintaining a constant and repeatable pressure (Fig. 7.5).

Fractional RF treatment involves the deep layers of the dermis and epidermis. The skin generates in a totally natural manner without pain, thus making an immediate return to a normal social life possible.

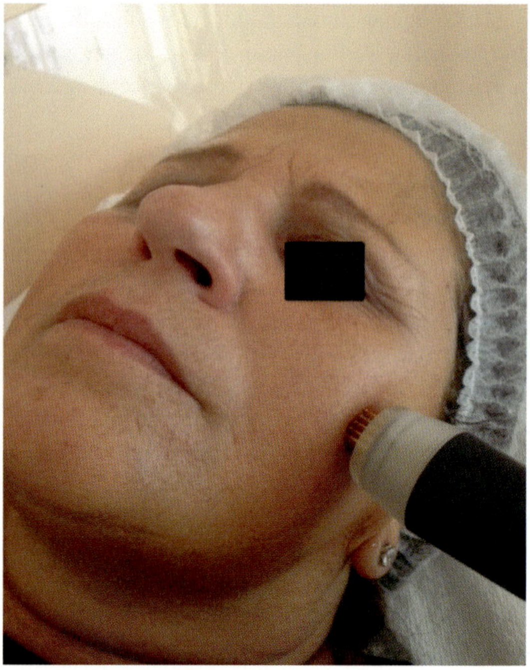

Fig. 7.5 Fractional RF handset

Indications for fractional RF are facial and body skin laxity, acne scars, wrinkles at all depths, and stretch marks [10].

7.10 Technique

The handset is applied to each area to be treated only once.

A few minutes after treatment an erythema appears over the treated area, which then disappears in the course of a few hours (Fig. 7.6).

It is advisable to apply a lenitive mask after treatment followed by a repair cream together with a strong sunscreen.

Three or four treatments are advisable, with a 2-week period separating each treatment.

After the initial treatment cycle, monthly maintenance treatments are advisable.

Visible results are evident after 4 months.

Two treatment cycles per year are advisable.

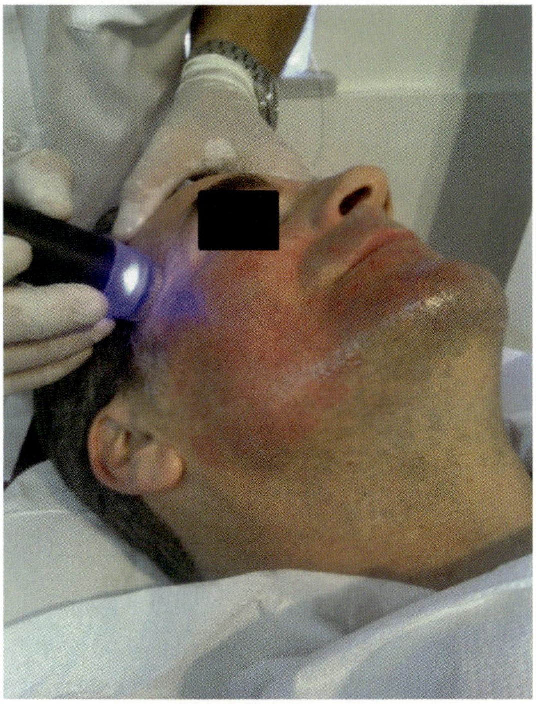

Fig. 7.6 Postinflammatory erythema treatment

7.11 Side Effects

Given that burn points that heal in 5–7 days are a possible collateral effect of fractional RF treatment, it is advisable to apply an antibiotic ointment and hyaluronic acid cream.

In addition, the application of a strong sunscreen is necessary for at least 1 month after treatment.

References

1. Beasley KL, Weiss RA (2014) Radiofrequency in cosmetic dermatology. Dermatol Clin 32(1):79–90
2. Sadick NS, Nassar AH, Dorizas AS, Alexiades-Armenakas M (2014) Bipolar and multipolar radiofrequency. Dermatol Surg 40(Suppl 12):S174–S179
3. Weiss RA, Weiss MA, Munavalli G, Beasley KL (2006) Monopolar radiofrequency facial tightening: a retrospective analysis of efficacy and safety in over 600 treatments. J Drugs Dermatol 5(8):707–712
4. Kassim AT, Goldberg DJ (2013) Assessment of the safety and efficacy of a bipolar multi-frequency radiofrequency device in the treatment of skin laxity. J Cosmet Laser Ther 15(2):114–117
5. Wollina U (2011) Treatment of facial skin laxity by a new monopolar radiofrequency device. J Cutan Aesthet Surg 4(1):7–11
6. Krueger N, Sadick NS (2013) New generation radiofrequency technology. Cutis 91(1):39–46
7. Alexiades AM, Rosenberg D, Renton B, Dovr J, Arndt K (2010) Blinded, randomized, quantitative grading comparison of minimally invasive fractional radiofrequency and surgical face lift to treat skin laxity. Arch Dermatol 146(4):396–405
8. Hruza G, Taub AF, Collier SL, Mulholland SR (2011) Skin rejuvenation. Dermatol Ther 24(1):41–53
9. Hantash BM, Renton B, Berkowitz RL, Stridde BC, Newman J (2009) Pilot clinical study of a novel minimally invasive bipolar microneedle radiofrequency device. Lasers Surg Med 41(2):87–95
10. Levenberg A, Gat A, Branchet M et al (2012) Treatment of wrinkles and acne scars using the tri-fractional, a novel fractional radiofrequency technology—Clinical and histological results. J Cosmet Dermatol Sci Appl 2(3):117–125

Biorevitalization and Combination Techniques

8

Maria Pia De Padova and Anna Masarà

Biorejuvenation is a common term to indicate mesotherapy for skin rejuvenation (also called biorevitalization or mesolift). It's a technique used to rejuvenate and tone the skin by means of an injection in the superficial dermis of suitable products, perfectly biocompatible and totally absorbable.

The goal of this technique is to increase the biosynthetic capacity of fibroblasts, inducing the reconstruction of an optimal physiologic environment, facilitating interaction between cells, and increasing collagen, elastin, and hyaluronic acid (HA) production.

The desired final effect is a firm, bright, and moisturized skin (Fig. 8.1).

8.1 Introduction

Chronoaging is responsible of clinical and histologic changes because of the intrinsic aging, like alterations in skin texture, elasticity, pigmentation, and modifications of subcutaneous tissue and the vascular system [1, 2]. Clinically, aged skin is characterized as thin, dry, and pale, with noticeable wrinkles and decreased elasticity [3]. Histologically, the epidermis becomes atrophic; there' s the accumulation of elastotic material in the papillary and mid-dermis, a process known as "solar elastosis," and quantitative changes in collagen, which are reflected in a decline in biosynthesis and content. The degree of changes is genetically determined and so different in each individual. Chronoaging can be worsened by cumulative environmental damages, such as chronic UV exposure (photodamage), pollution, and smoking. The effect of photodamage, termed "photoaging," is characterized by wrinkles, shallowness, laxity, patchy pigmentation, and rough-textured skin that histologically are signs of hyperplasia or atrophy. Dermal features include elastosis, degeneration of collagen, anchoring fibrils, and dilated and twisted blood vessels.

UV exposure activates free radicals and matrix-degrading metalloproteinase enzymes, including collagenase [4, 5].

8.1.1 When and Where

The mesotherapy can be performed in cases of mild to moderate chronoaging and photoaging, aging prevention, and preparation to sun exposure and smokers.

The treatment is indicated for both young skin, yet elastic and vital, to reduce the physiological

M.P. De Padova
Department of Dermatology, University of Bologna, Bologna 40138, Italy
e-mail: mdepadova@gmail.com

A. Masarà
Division of Clinical Dermatology, Department of Clinical Medicine and Surgery, University of Naples Federico II, Via Sergio Pansini 5, Naples 80133, Italy

G. Fabbrocini et al. (eds.), *Nonsurgical Lip and Eye Rejuvenation Techniques*,
DOI 10.1007/978-3-319-23270-6_8

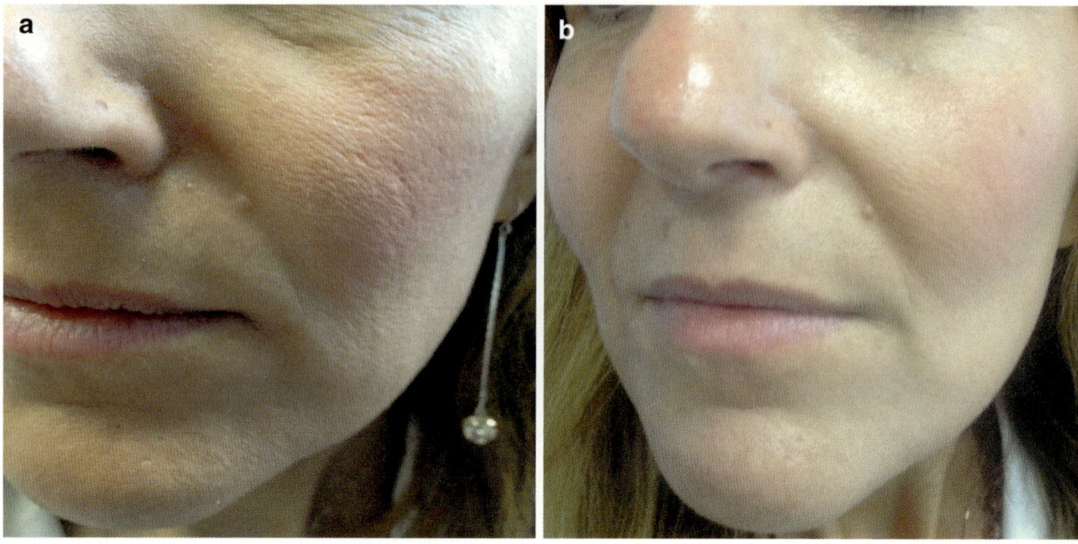

Fig. 8.1 (**a**) Left side before treatment, (**b**) left side after treatment

aging process, because the used substances ensure deep hydration that delays the onset of age-related imperfections and counteracts oxidative damage that is caused by environmental factors and exposure to sunlight, and for mature skins, to reduce the signs of chronoaging by reactivating the cellular functionality.

Areas of application are [6, 7]:

- Face
- Neck
- Low neckline (décolleté)
- Dorsum of hands
- Belly
- Inner surface of arms and legs

8.1.2 Functions

Biorevitalization performs three main functions:

- Restructuring: it promotes cell turnover and the production of collagen, elastin, and hyaluronic acid.
- Antioxidant: it protects the skin from free radicals that are released as a result of environmental factors and solar radiation.
- Moisturizing: it promotes the rapid recall of water in the tissues.

It allows to obtain substantial improvements in terms of (Fig. 8.2):

- Firmness and elasticity of the skin
- Brightness
- Reduction of fine lines

8.1.3 Advantages vs. Contraindications and Side Effects

Mesotherapy injection is a minimally painful procedure and requires no anesthesia.

The sessions are performed in cycles and do not involve side effects except in rare cases, a slight temporary redness or a small bruise in the area due to the trauma of the needle insertion, which tends to disappear spontaneously in 2–3 days, and usual activities can be resumed immediately.

The main advantages/disadvantages and contraindications are listed in Table 8.1.

8.2 Available Products

The desired effect – firm, bright, moisturized skin (Fig. 8.3) – can be achieved by microinjections in the superficial dermis of products containing

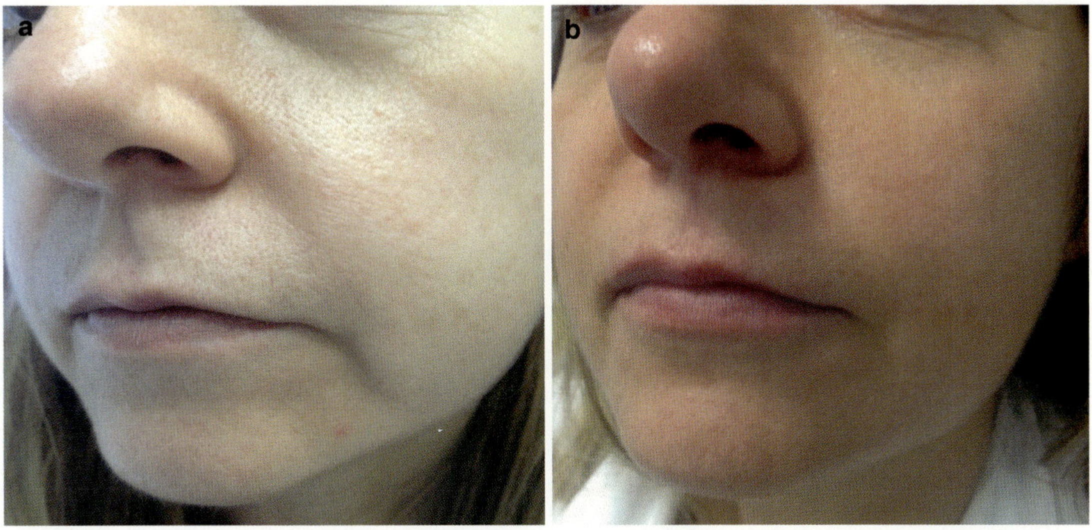

Fig. 8.2 (**a**) Left side before treatment, (**b**) left side after treatment

Table 8.1 Advantages, contraindications and disadvantage of biorejuvenation techniques

Advantages	Contraindications	Disadvantages
Easy to perform Low pain No necessity for skin tests Limited side effects No downtime or recovery time Suitable for every skin phototype	Allergy to the ingredients injected History of hypertrophic scars Bleeding abnormalities and/or anticoagulant therapy Pregnancy/breast-feeding Autoimmune disorder (lupus, scleroderma) Epilepsy Diabetes Active phase of Herpes simplex virus 1 infections Bacterial infections Inflammatory skin disorders	Only for mild-moderate aging Mild erythema, slight itching/burning sensation 5 min after injections Small hematomas Possibility of allergic reactions Lack of controlled clinical trial Lack of guidelines according to the evidence-based medicine

only one active ingredient or "cocktails" of different compounds that are biocompatible and absorbable.

The products available in mesotherapy for skin rejuvenation are:

- Hyaluronic acid alone (1.35–3 %)
- Hyaluronic acid 0.2, 1, or 3 % plus other active ingredients like vitamins, amino acids, minerals, coenzymes, nucleic acids, and β-glucan
- Macromolecules
- Organic silicium
- Autologous cultured fibroblasts
- Growth factors
- Homeopathic products

The most frequently used substance is natural non-cross-linked hyaluronic acid (Fig. 8.4). Chemically HA is the major nonsulfated glycosaminoglycan of the connective tissue scaffold, synthesized by fibroblasts within the cell membrane and then released in the extracellular space [8, 9]. Interestingly, the injection of simple HA not only provides enrichment of one of the main ECM compounds and deep hydration of the skin but also stimulates fibroblasts, acting on specific receptors (CD44, RHAMM, and ICAM-1)4 to synthesize new scaffold compounds [10].

One gram of HA can bind up to 6 L of water. This means that the higher the percentage of HA is in a composition (milligrams of HA per milliliter),

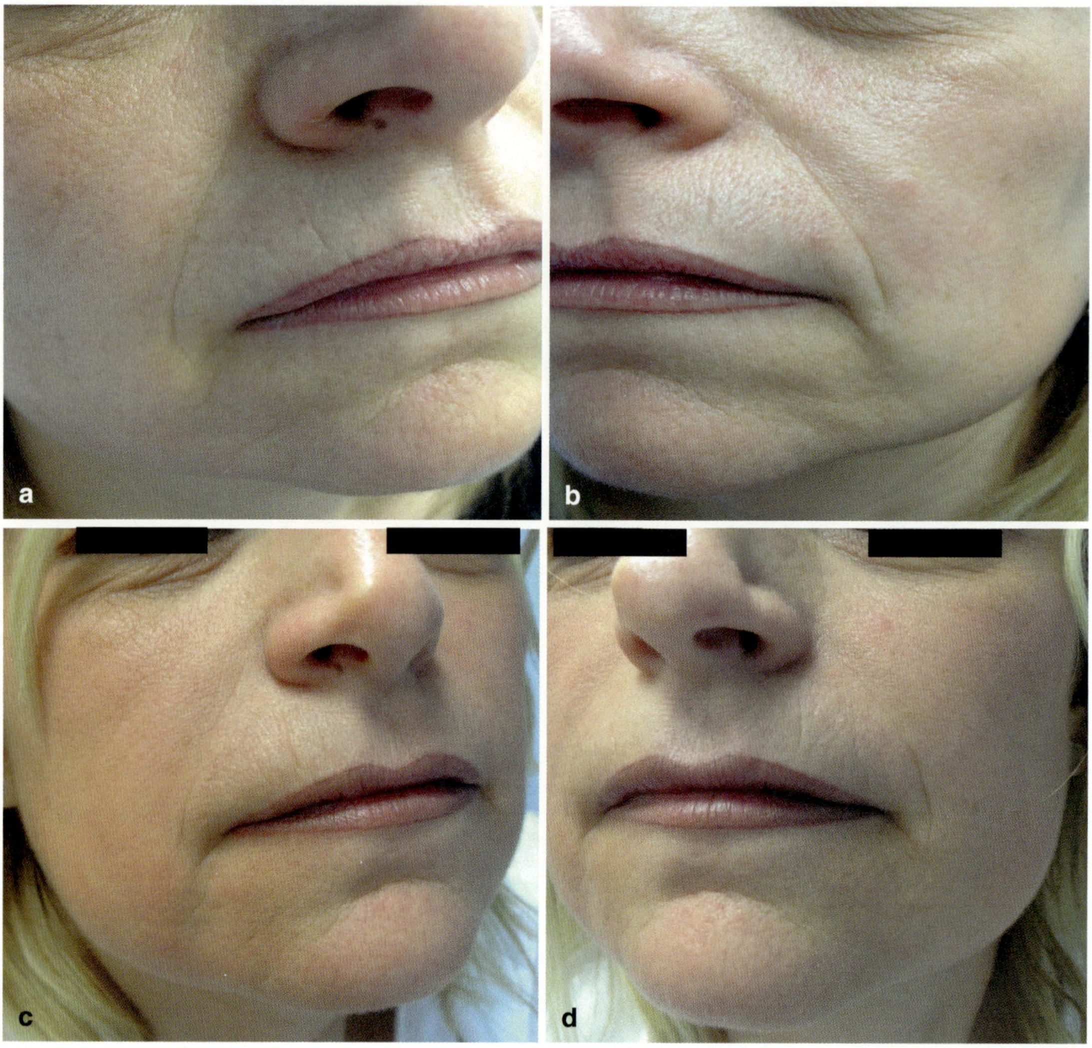

Fig. 8.3 (**a**) Right side before treatment, (**b**) left side before treatment, (**c**) right side after treatment, (**d**) left side after treatment

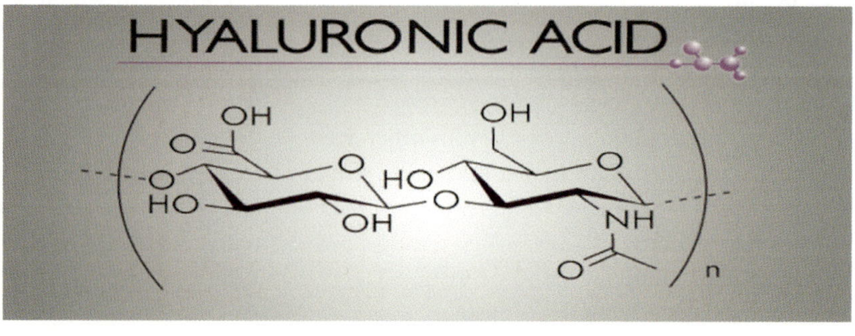

Fig. 8.4 Chemical structure of HA

the higher its capacity to retain water will be. Derived either from rooster combs or from bacterial fermentation, HA has no species specificity, and the risk of a hypersensitivity reaction is so low that skin testing is unnecessary. Hyaluronic acid used in mesotherapy is not cross-linked, it is not much stable, it is very fluid, and it has a short half-life, even shorter than the one used in fillers.

The great versatility of biorevitalization lies in the different biological effects of the injected active substances. The synergy of different functional ingredients can treat skin in a more complete way, acting on various age-related marks caused by both intrinsic and extrinsic aging factors, with a preventive and curative action [11].

In a mesotherapy cocktail, vitamins are the most important active component:

- Vitamin A regulates the epidermis turnover and it is an antidrying agent.
- The "vitamin B complex"—usually indicating a group of vitamins that includes vitamin B1 (thiamine), B2 (riboflavin), B3 (niacin), B5 (pantothenate), B6 (pyridoxine), B9 (folic acid), and B12 (cyancobalamin)— includes coenzymes involved in several metabolic processes that help the scavenging of free radicals.
- Vitamin C is a well-known antioxidant and it induces collagen synthesis.
- Vitamin E is an antioxidant and moisturizer
- Vitamin K has an effect on microcirculation
- Vitamin D, vitamin H (biotin), vitamin B10, and vitamin I (inositol) are important too.

The amino acids build polypeptides constituting the matrix of cell architecture. Sodium, potassium, calcium, and magnesium act as catalysts in numerous cell functions. Coenzymes are nonprotein organic components that help the enzymes in their catalytic function. They are "activators" of biochemical reactions and help the dermis turnover. DNA and RNA are bound to proteins, and they give information for the regulation of protein synthesis. β-Glucan acts as a free radical scavenger. Polynucleotidic macromolecules favor skin hydration increasing water retention; they act as scavengers of free radicals, and they enhance the physiologic activity of fibroblasts [12].

8.3 Infiltration Techniques

There are different injection techniques in the superficial dermis (Fig. 8.5) that can be performed always keeping the needle with an inclination of 45°:

- Picotage: it's more useful in younger people who want to prevent skin aging due to sun exposure and tanning sunbeds. It consists of microinjections, is very superficial, is practically painless, and is spaced at 2 mm, and the needle penetrates the treated area at 2–2.5 mm. During the procedure the physician maintains a constant pressure on the plunger. The mostly cured areas are the face, neck, decollete, and less frequently hands (Fig. 8.6).
- Cross-linking: it is recommended for the prevention and treatment of skin aging (patients with a more advanced stage of chronoaging, compared to the prior technique). It consists in performing intradermal linear infiltrations

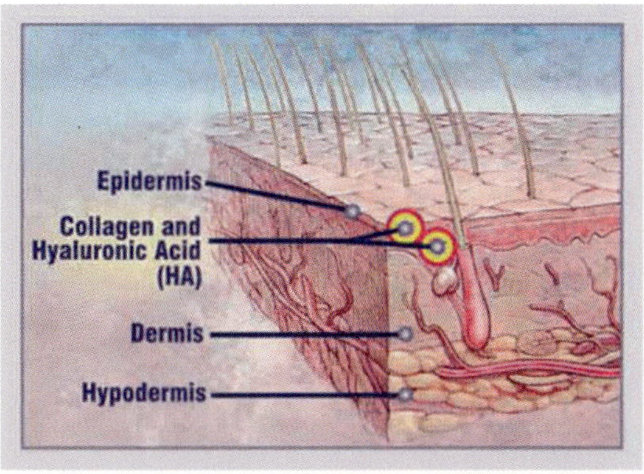

Fig. 8.5 Epidermal layers – where to inject HA

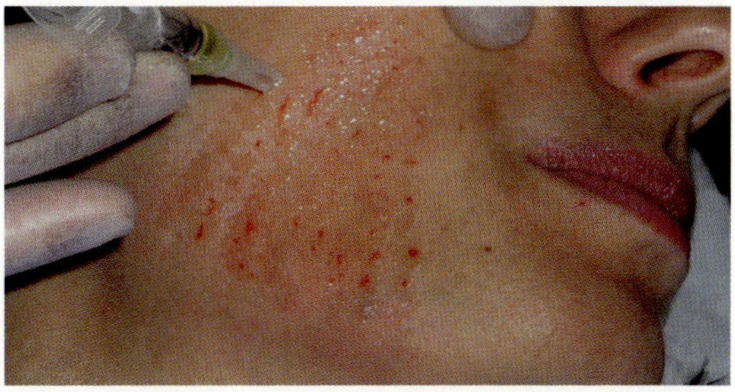

Fig. 8.6 Picotage technique

with a complete penetration of needle vertically and horizontally, spaced at 1 cm, to form a grid. The product is injected during the extraction of the needle from skin dermis.

- Linear threading: either vertical or horizontal injections are performed. Vertical injections are useful to prepare the nose-labial and glabellar wrinkles 10–15 days before injecting dermal fillers and botulinum toxin. Horizontal injections are useful in treating neck wrinkles.

To reduce the burning sensation, the physician can apply anesthetic cream 1 h before the treatment. It is recommended to avoid injecting products containing also vitamin C.

After the treatment, a gentle massage with a vitamin K cream can be given. The procedure generally takes about 20 min, but it may vary depending on the treated area. Sun and smoking avoidance are recommended for the next 48 h. There is no downtime or recovery time with this procedure (Fig. 8.7). Treatments should be done once every 2 weeks for 3–4 weeks, then once a month for 3–4 months. The results are maintained by touch-up treatments once or twice a year. This protocol may vary according to the patient's age, clinical presentation at first visit, and response to initial treatments. Typically, 2–3 treatments are necessary to see some results, even if the brightness is visible after the first treatment, due to a vascular stimulus by the microinjections. The number of treatments can vary from patient to patient and depends on the treated area and patient's expectations. It is important to remember that mesotherapy is not a filling technique, but it permits the rejuvenation of the skin by increasing its hydration and by reconstructing an optimal physiologic environment for the fibroblasts.

8.4 Combination Techniques

Sessions of biorevitalization can be performed in combination with other treatments, surgical or not, such as peeling, face lift, eyelid surgery, fillers, infiltration of botulinum toxin, and laser treatments. Biorejuvenation can be used to prepare the skin 2 weeks before the injections of other products.

This is because biorevitalization works improving globally the skin unlike other treatments that act more locally.

Among these are:

Peelings The purposes of chemical peelings are to erase shallower wrinkles, restore tone, give freshness and radiance to the face, eliminate dark spots and scars, etc. Peelings represent an accelerated form of exfoliation induced and controlled by the use of one or more caustic chemicals applied to the skin.

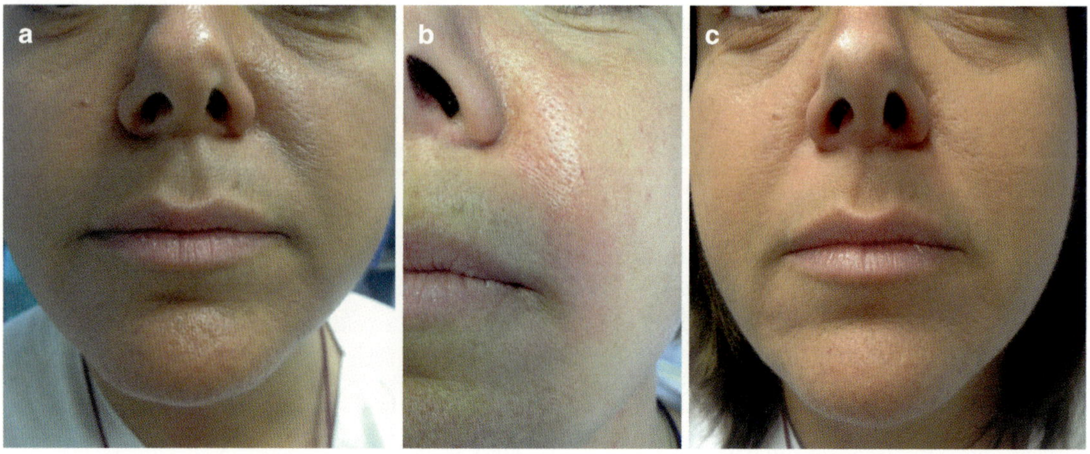

Fig. 8.7 (**a**) Front side before treatment, (**b**) front side during treatment, (**c**) front side after treatment

Depending on the agent used, its concentration and the time of application, these preparations cause a partial or total programmed destruction of the epidermis.

The effect is the stimulation of cellular turnover through the removal of the horny layer and, simultaneously, the induction of the synthesis of new collagen in the dermis.

The result is the replacement of old tissue with a healthier and less corrupt one.

There are three broad categories of peelings which act in different ways depending on the depth of penetration of exfoliating: the superficial peelings, middle, and deep ones.

- Superficial peelings: they allow the immediate resumption of work and social activities. The treatment is ambulatory and is composed of cycles with more spaced sessions (intervals of 7–15 days apart), also can be performed twice a year, for example, glycolic acid peels, salicylic acid peels, and retinoic acid peels.
- Medium or deep peelings: they are suitable for individuals with badly damaged skin by the chrono- and photoaging or for skin that presents wrinkles and rather deep furrows and spread discoloration. This kind of peelings requires a recovery period at home during which skin exfoliation occurs and is much stronger than after a superficial peel. An example is trichloroacetic acid peel that may be superficial, medium, or deep depending on the concentration of acid used.

The results depend on the type of peel made; for example, a deep one has a definite result and only one treatment is required, while the medium peel and the superficial ones are less invasive, but allow less lasting results and therefore need to be repeated regularly.

All types of peels require careful and meticulous aftercare treatment, consisting basically in no sun exposure until healing has occurred (or cycle ended) and the use of sunscreens, emollient creams, and products based on alpha-hydroxy acids for home use.

Fillers In cosmetic medicine "fillers" are materials that are injected into the dermis or subcutaneous tissue in order to fill a depression or increase the volume. They may be transient, when their cosmetic-clinical effect ceases after some time, and permanent, remain where injected, for life. Hyaluronic acid gel fillers are now the most used and safe and they are completely resorbable. Another widely used material for particular anatomical areas is the calcium hydroxyapatite (this filler is usually composed for 30 % of microspheres of synthetic calcium hydroxylapatite (CaHA) and 70 % from an aqueous gel solution).

No one of the so-called "permanent" fillers has been approved by the FDA.

The permanent fillers (non-resorbable) were basically created for the need to make a correction very durable. Unfortunately, experience has shown that very often this intent is illusory, not because the substance does not remain long in the skin tissue but because it has its own weight and density that lead it slowly to migrate from the implant site following the force of gravity. So there is a progressive appearance of the filler in a different region from the injection site.

More serious is the frequent appearance, years after the implantation, of very serious local inflammation, abscesses, and granulomas or thick areas of fibrotic tissue that follow for months and years tormenting the patient's appearance, his/her health, and his/her normal social life.

The best known is the liquid silicone, however, banned in Italy by ministerial decree in 1993.

Fillers can be used to treat facial wrinkles, after a restrictive and low-calorie diet or after a prolonged period of illness, when we see the area of the cheeks gaunt; very satisfactory is the installation of fillers in the lips for young women and even older women to give volume and firmness to the face, to redesign the profile of the nose (rinofiller), for reconstructive purposes, depressed scars, results of acne and chicken pox, asymmetries, and facial atrophies, and in all cases where we want to take care of ourselves from the aesthetic point of view.

The hyaluronic acid gel is injected into the dermis with thin needles of different diameters and lengths depending on the viscosity of hyaluronic acid and the area to be treated or with ago cannulae.

The aesthetic result is very natural and the absorption of the substance is gradual. The material is fully resorbable, and its duration is very satisfactory, generally 8–12 months or more, depending on the areas treated, the individual characteristics, and lifestyle.

After the injections mild redness is frequent, but disappears in a matter of hours without a trace, and variably, a slight swelling. You could also have the formation of some bruisings or hematomas (caused by rupture of a small capillary during injection) which resolves spontaneously and in a short time with the local application of creams based on vitamin lactoferrin.

It is not advisable to expose the treated area directly to the sun or sunlamps. It is also advisable not to rub and massage the site for 24 h and not to apply any makeup for at least 3–4 h after.

Botulinum toxin The botulinum toxin is a protein produced by the bacterium *Clostridium botulinum* that, by blocking the transmission of nerve impulses, reduces muscle contraction. Only a fraction of the toxin (type A) is used in aesthetic medicine, purified and diluted in saline.

After a careful study of facial expressions of the person, the physician practices many small injections in the chosen areas, paralyzing the underlying muscles producing a "lifting effect."

The areas of choice of Botox are wrinkles around the eye and the eyebrow, but also the wrinkles of the forehead. In other words, the Botox is suitable for correcting the wrinkles arising from facial muscle movements (facial expression muscles). The effect is temporary, 4–6 months, and starts a week after the infiltration. The microinjections of botulinum toxin which, of course, are not toxic are not painful and do not cause swelling.

There is a theoretical risk of a hypersensitivity reaction to the product itself or to the additives contained in it, which in any case is directly proportional to the amount administered.

In particular, the current formulations of botulinum are contraindicated for people with allergies to milk, because they are used as a preservative albumin. Furthermore, the botulinum toxin is not recommended during pregnancy and lactation. The "mesobotulino" is one of the most interesting methods in the field of rejuvenation. It consists of microinjections of botulinum toxin in very diluted form, with a cocktail of amino acids and vitamins that aim to correct the oval of the face, cheeks, and sagging chin and neck profile.

The action of vitamins, combined with the paralyzing botulinum, smoothes the skin and gives a lifting effect. The treatment is safe and provides only a slight annoyance given by injections.

Laser Fractional CO_2 laser is the first choice for sun damage, wrinkles, and texture because it eliminates the superficial layers of the skin, stimulating at the same time, the contraction of the fibers of collagen, and elastin in the dermis. The fractional CO_2 laser can shrink the skin and reduce the appearance of fine wrinkles and large pores; it can act effectively even on acne scars and skin discoloration.

It practically eliminates the superficial layer of the skin and, at the same time, strongly stimulates the deep layers so as to have an intense process of rejuvenation or tissue repair. The use of the fractionated scanner instead of the previous type "ablative" greatly reduces the down time (time to return to social life after the treatment) because the microareas treated are spaced with free areas with the result of a more rapid healing.

References

1. Yaar M, Gilchrest BA (2007) Photoaging: mechanism, prevention and therapy. Br J Dermatol 157:874–887, PubMed: 17711532
2. Farage MA, Miller KW, Berardesca E et al (2009) Clinical implications of aging skin: cutaneous disorders in the elderly. Am J Clin Dermatol 10:73–86, PubMed: 19222248
3. Makrantonaki E, Zouboulis CC (2007) Molecular mechanisms of skin aging: state of the art. Ann N Y Acad Sci 1119:40–50, PubMed: 18056953
4. Tosti A, Grimes PE, De Padova MP (2006) Atlas of chemical peels. Springer, Berlin
5. Rabe JH, Mamelak AJ, McElgunn PJS et al (2006) Photo aging: mechanisms and repair. J Am Acad Dermatol 55:1–19

6. Cavallini M (2004) Biorevitalization and cosmetic surgery of the face: synergies of action. J Appl Cosmetol 22:125–132
7. De Padova MP, Bellavista S, Iorizzo M et al (2006) A new option for hand rejuvenation. Pract Dermatol 8:12–15
8. Andre P (2004) Hyaluronic acid and its use as a "rejuvenation" agent in cosmetic dermatology. Semin Cutan Med Surg 23:218–222
9. Monheit G, Coleman KM (2006) Hyaluronic acid fillers. Dermatol Ther 19:141–150
10. Ghersetich I (1997) Management of aging skin. J Eur Acad Dermatol Venereol 9:51
11. Sparavigna A, Tenconi B, De Ponti I (2015) Antiaging, photoprotective, and brightening activity in biorevitalization: a new solution for aging skin. Clin Cosmet Investig Dermatol 8:57–65
12. Iorizzo M, De Padova MP, Tosti A (2008) Biorejuvenation: theory and practice. Clin Dermatol 26:177–181

Laser for Periorbital Rejuvenation 9

Julia P. Neckman, Jeremy Brauer,
and Roy G. Geronemus

9.1 Introduction

Since Albert Einstein first developed the concept of laser radiation, physicians have used lasers along with other components of the electromagnetic spectrum in a variety of medical and cosmetic applications. Appreciation of the physics behind lasers provides a foundation for understanding its applications. The electromagnetic spectrum comprises radiation energy spanning short gamma waves to long radio waves, and in between it includes x-rays, ultraviolet radiation, visible light, infrared light, and microwaves. If sufficient electromagnetic radiation is absorbed by resting atoms, their electrons are stimulated to excited states. When these electrons eventually return to resting states, the atom releases the same absorbed energy at its wavelength in a process known as "spontaneous emission."

Spontaneous emission may be hastened, or stimulated, when an excited atom is irradiated a second time with the same wavelength used to excite it originally. The second hit may come from a new source of energy or from spontaneous emissions of nearby atoms. As a result, if atoms are concentrated in a particular medium and confined within a reflective charged cavity, emissions may become markedly amplified because of the interaction of the spontaneous emissions and the surrounding stimulated atoms.

Maiman was the first to demonstrate Einstein's theories of stimulated emission using visible light [1]. This led to his coining the now familiar acronym LASER, which stands for light amplification by stimulated emission of radiation. Although light technically refers to the visible spectrum, all laser emissions, whether in the visible spectrum or not, are generally referred to as laser light. The wavelength of the laser light is dependent on the medium of the reflective charged cavity. In the original Maiman study, ruby crystal made up the medium, but since that time several other mediums, such as alexandrite, potassium titanyl phosphate (KTP), and others, have been used to generate other wavelengths in medicine.

Laser light is monochromatic, coherent, and collimated. Monochromaticity results from its consisting of one wavelength. Coherence refers to light waves that travel in phase, both in time and space, and collimation relates to the parallel nature and lack of divergence of the light waves.

Use of lasers by physicians was revolutionized by the concept of selective photothermolysis [2]. Essentially, selective photothermolysis takes advantage of the heterogeneous absorption spectra of anatomic structures, particularly chromophores

J.P. Neckman, MD (✉) • J. Brauer, MD
R.G. Geronemus, MD (✉)
Laser & Skin Surgery of New York, New York University Medical Center,
317 East 34th Street, New York, NY 10016, USA
e-mail: jneckman@laserskinsurgery.com; rgeronemus@laserskinsurgery.com

G. Fabbrocini et al. (eds.), *Nonsurgical Lip and Eye Rejuvenation Techniques*,
DOI 10.1007/978-3-319-23270-6_9

such as melanin, hemoglobin, and water. Laser light energy absorbed by a target chromophore is converted primarily to heat, destroying the chromophore itself and the surrounding cell. The heat created at the site of the target chromophore may dissipate to surrounding cells, causing their destruction. The preferential absorption of these structures for different wavelengths permits their targeted ablation, coagulation, or thermal damage with important preservation of surrounding structures.

Successful laser use, however, relies on more than wavelength and target. Training, experience, and management of settings such as fluence, spot size, and pulse width are critical to safe and effective clinical outcomes. Fluence is a measure of the laser's energy in joules per centimeter squared. Spot size is clinically important since larger spot sizes may cause peripheral damage around small targets. It also results in deeper penetration of the laser's effects, but with more scatter. Pulse width is a measure of laser exposure time and is clinically relevant because of its relationship to thermal relaxation time (TRT). For a given tissue target, TRT is the time required to lose half of its heat. If the pulse width is longer than the TRT, less ablation and more surrounding damage in the form of coagulative necrosis occurs.

The objective of this chapter is to review common applications of lasers for the treatment of periorbital concerns. These applications are broad and include resurfacing as well as the elimination of unwanted vascular and pigmented lesions. Although some of the technologies discussed in this chapter now serve as newer tools in traditional surgery, such as lasers in place of scalpels for making incisions, the following discussion will primarily concentrate on the role of the technologies when employed as the primary therapeutic intervention.

9.2 Periorbital Photodamage and Rejuvenation

Noninvasive and minimally invasive treatments for periorbital photodamage and rejuvenation have grown markedly in recent years. Today's strategies commonly employ resurfacing lasers in addition to the related technology of radio frequency, which is discussed in a separate chapter. These technologies have shown effectiveness improving skin laxity, rhytides, scars, and more recently premalignant changes of the skin, namely, actinic keratoses [3].

Originally, resurfacing lasers were nonfractional and fully ablative carbon dioxide lasers. Though they delivered impressive results, they came with substantial risks, particularly for scarring and hypopigmentation. With regard to periorbital treatments, scarring could further lead to ectropion, entropion, and epiphora. Nonfractional, non-ablative laser alternatives followed, which were safer, but delivered less impressive results. With the advent of fractional lasers, meaningful rejuvenation became achievable with far less risk [4]. The seminal concept of fractional laser delivery was first described in 2004 and has since been applied to non-ablative and ablative devices [5]. In essence, a fractional system delivers laser in a pixilated pattern, creating zones of injury surrounded by areas of unaffected skin. Evidence for the efficacy and safety of these systems for periorbital treatment is now supported throughout the literature.

A retrospective study of 31 patients treated with a non-ablative fractional resurfacing laser to the upper and lower eyelids was evaluated for changes in eyelid tightening and eyelid aperture [6]. The laser consisted of a fractionated 1,550 nm erbium-doped fiber laser and was delivered over 3–7 treatment sessions. All patients achieved eyelid tightening, without any concerning adverse effects or downtime. Just over half, specifically 55.9 %, also achieved increase in eyelid aperture. Improvement in eyelid tightening and aperture can be seen in Fig. 9.1. Ablative systems have also been formally evaluated around the eye. A prospective study of 15 patients evaluated the effect of an ablative fractional carbon dioxide laser resurfacing treatment for laxity of the eyelid and periorbital skin [7]. Investigators found a 53.1 % improvement in rhytides and 42.0 % improvement in skin redundancy. The only adverse effects reported in the study were two patients experiencing post-inflammatory hyperpigmentation that resolved after 3 months with hydroquinone and sunscreen. While that study did not report any serious adverse effects, ectropion has been

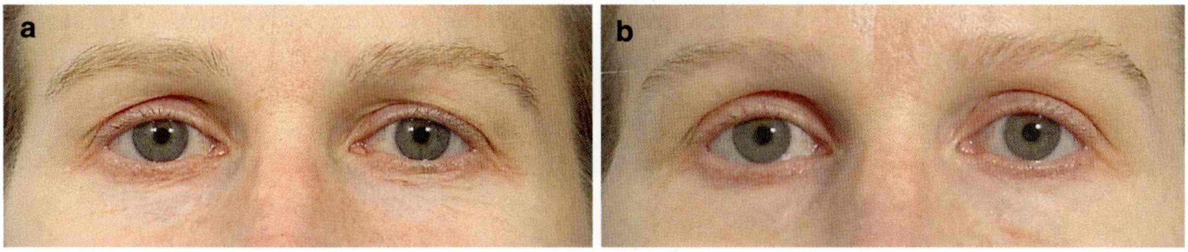

Fig. 9.1 Baseline (**a**) and follow-up (**b**) photos 1 month after third non-ablative fractional photothermolysis treatments of upper and lower eyelids from the lid margin to the orbital rims

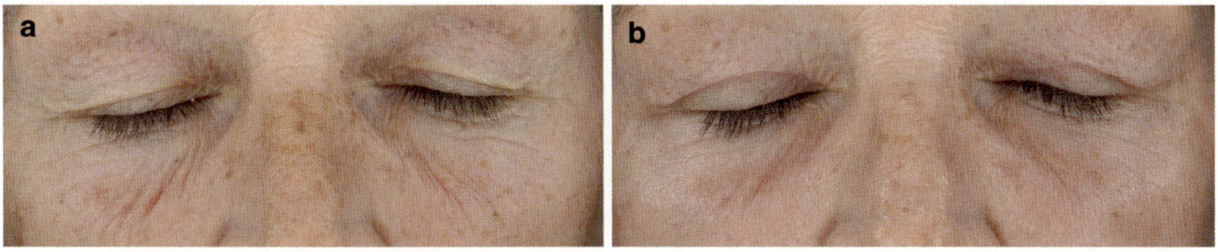

Fig. 9.2 Subject treated with fractional CO_2 laser for periorbital rejuvenation with noted reduction of upper eyelid hooding, rhytides, infraorbital tear trough discoloration, and crepe-like skin. (**a**) Baseline. (**b**) Three months post procedure

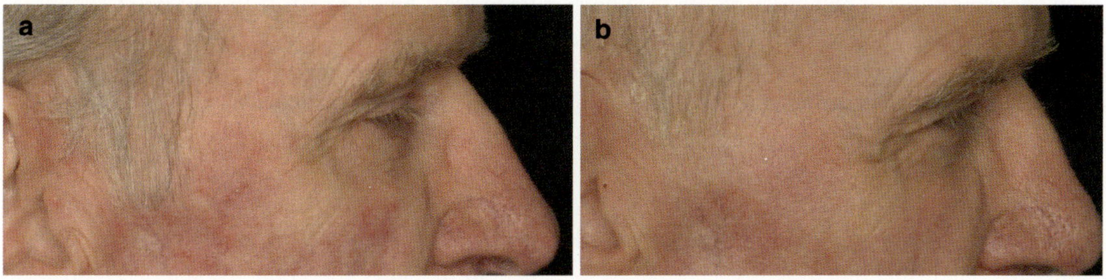

Fig. 9.3 Facial actinic keratosis (AK) and photodamage before treatment (**a**) and reduction in facial AK and photodamage at 6 months after fourth treatment session with fractionated 1927 nm laser (**b**)

reported following ablative fractional carbon dioxide resurfacing on the lower eyelids [8]. Overall, improvement in skin laxity and fine rhytides with fractional carbon dioxide resurfacing can be seen in Fig. 9.2. Surgeons should notably be aware of the value of these rejuvenating laser systems for the improvement of surgical scars, since their efficacy has been shown for a variety of scar types [9, 10]. One study objectively and quantifiably examined the effect of ablative fractional carbon dioxide laser resurfacing of 19 atrophic scars resulting from surgery or trauma [11]. Subjects were treated 3 times and followed for 6 months. Subjective assessment of treated scars both by investigators and patients found improvement in skin texture. These findings were confirmed by optical tomographic analysis that quantifiably demonstrated a 38.0 % mean reduction of volume and 35.6 % mean reduction of scar depth.

In addition to the cosmetic enhancements achievable with these technologies, the medical value should not be underestimated. Periorbital skin cancers commonly challenge ophthalmic and dermatologic surgeons. Laser resurfacing has recently been shown to be valuable against premalignant changes, namely, actinic keratoses, which may serve as precursors to squamous cell carcinoma [3]. The mechanism of therapy is not yet understood, but the clinical response is evident, as in Fig. 9.3. The authors of this chapter frequently employ a non-ablative fractional thulium 1927 nm laser as field treatment to reduce actinic keratoses over the face, including periorbital skin.

9.3 Pigmented Concerns of the Periorbital Skin

Several pigmented concerns of the periorbital skin respond to laser therapy. Commonly used lasers for pigment include the ruby, alexandrite, diode, and neodymium-doped yttrium aluminum garnet (Nd:YAG), as their wavelengths can target melanin. These lasers effectively treat periorbital lesions such as ephelides, lentigines, café au lait spots, nevi of Ota, congenital melanocytic nevi, and tattoos.

Ephelides, also known as freckles, arise on sun-exposed areas as well-defined circular or oval hyperpigmented macules of just a few millimeters. While not precancerous themselves, high concentrations of ephelides on the face have been associated with genetic variations in the melanocortin 1 receptor (MC1R) [12]. MC1R gene variants are also associated with fair skin, red hair, and melanoma and nonmelanoma skin cancer. On pathology, an ephelide demonstrates normal epidermal configuration, but tends to have larger melanocytes in the basal layer with additional dendritic branching.

Lentigines may be characterized as simple or solar and may involve mucosal surfaces, unlike ephelides. Simple lentigines arise at an earlier age and in any location in contrast to solar lentigines, which arise in adulthood and in sun-exposed areas. Solar lentigines appear with increasing age and are a sign of photodamage [13]. Lentigines are well-defined circular or oval hyperpigmented macules and tend to be slightly darker and larger. Lentigines display elongated rete ridges on pathology, with more numerous melanocytes than typical skin. The solar lentigo has rete ridges that are more uniform and clubbed in appearance when compared to the rete ridges of a simple lentigo [14]. Lentigines are associated with several genetic syndromes, including LEOPARD, which also demonstrates ocular hypertelorism, and Peutz-Jeghers syndrome, which often includes periorbital and conjunctival lentigines.

Ephelides and lentigines are treated similarly and effectively with lasers targeting pigment. A study of 34 pigmented lesions, including lentigines, was performed using the Q-switched ruby laser at settings of 694 nm, 40 ns pulse duration, and 4.5 and/or 7.5 J/cm^2 [15]. Substantial clearing was appreciated in the lentigines with just one treatment at either fluence. Long-term follow-up reveals efficacy in the majority of patients with lentigines. In another study of 10 patients with solar lentigines treated once or twice with the Q-switched ruby laser, 77 % demonstrated continued response at 10–21 months follow-up [16]. Photopigmentation was also found to be effectively treated with a 1927 nm non-ablative fractionated thulium laser. Treatment produced moderate to marked improvement in overall appearance and pigmentation with high patient satisfaction. The response to treatment was maintained at 1 and 3 months' follow-up [17].

Congential melanocytic nevi are collections of melanocytes presenting as flat or raised blue-brown lesions, with or without excess hair, and with an increased risk of melanoma when giant [18]. Intervention depends on risk of progression to melanoma, cosmetic disfigurement of the lesion, and complexity of removal. Laser therapy is helpful, but is also controversial based on questions of dysplastic effects of lasers on nevi and an increased risk of melanoma in some congenital nevi [19]. In addition, recurrence of lesion and color is not uncommon. One strategy for treatment is laser ablation. In a study of 13 patients with medium-sized congenital nevi, as much tissue as possible was excised, followed by erbium/YAG ablation of residua [20]. 83 % of patients were rated as having good to excellent results by the physician global assessment scale, and 77 % of patients reported good to excellent results at 4 months after treatment. Ablative lasers have also been successful in dark skin types [21]. Another approach involves pigment-specific lasers. In a study, 9 patients with medium-sized congenital nevi on the face or upper limbs were treated on average 9.6 times with a Q-switched ruby laser [22]. After treatment, 0–20 % of the lesions' color remained. However, 8 demonstrated slight repigmentation that responded to additional treatment. One lesion returned to its original color within a month of its final treatment and therefore was simply excised.

The periorbital pigment concerns in Asian skin can also pose a challenge in terms of correcting nevi. In one study, 7 small congenital nevi in 24 Korean patients were treated with Er:YAG laser followed by long-pulsed alexandrite laser at 1-month intervals. At 8 weeks after the final treatment, all treated nevi showed complete removal of pigmentation with only one recurrence of pigment after 6 months [23].

A nevus of Ota is a blue-brown patch that usually arises in infancy, around puberty, or in pregnancy. A favored site is the periorbital skin. Often the ipsilateral sclera shows blue-brown hyperpigmentation as well. Less commonly, other components of the eye can be affected, and, importantly, glaucoma may be seen in 10 % of those affected with nevus of Ota [24–26]. The pathology explains the blue hue to the skin, namely, a higher than normal concentration of dermal melanocytes. Laser treatment is effective against the cutaneous periorbital features of nevus of Ota and has been demonstrated with numerous lasers, including the alexandrite, ruby, and Nd:YAG [27–29]. Notably, these same laser measures are not safe as therapy for scleral involvement of the pigment. One study composed of 602 Chinese nevus of Ota patients found benefits with each additional treatment using a Q-switched alexandrite laser [30]. The study also found poorer response on the eyelid skin, which is referred to by some as the "panda sign." Another study of 119 nevus of Ota patients demonstrated a marked periorbital under-response [31]. They recommend discounting the traditional Tanino classification of nevus of Ota, which is based on clinical distribution, and instead adopting a system based on response to laser treatment.

In addition to the endogenous pigment concerns discussed above, exogenous pigment in the form of tattoos can be found periorbitally. The steady rise in the use of tattooed makeup around the eyes has been accompanied by, not surprisingly, an equally steady rise in the number of patients seeking periorbital tattoo removal. Common challenges to periorbital tattoo removal are preservation of hair follicles, as these tattoos are typically placed along the eyelash and eyebrow lines, and the avoidance of red, white, and beige/brown tattoo pigment since these pigments paradoxically darken with Q-switched laser treatment [32]. The phenomenon of paradoxical darkening is generally attributed to the reduction of ferric oxide ($Fe3+$) to ferrous oxide ($Fe2+$) in the pigment. Because of both challenges, periorbital tattoo removal is often accomplished with the careful use of a Q-switched Nd:YAG and/or an ablative laser, particularly a fractionated carbon dioxide laser. The Q-switched Nd:YAG allows for small spot sizes that better target fine eyelid tattoos and lessen risk of adjacent follicle damage [33]. An effective response to the Q-switched Nd:YAG can be seen in Fig. 9.4. Ablative lasers are incorporated when red, white, and beige/brown pigments are present, since these lasers do not cause paradoxical darkening [34]. These ablative lasers effectively clear tattoos via superficial tissue vaporization and subsequent

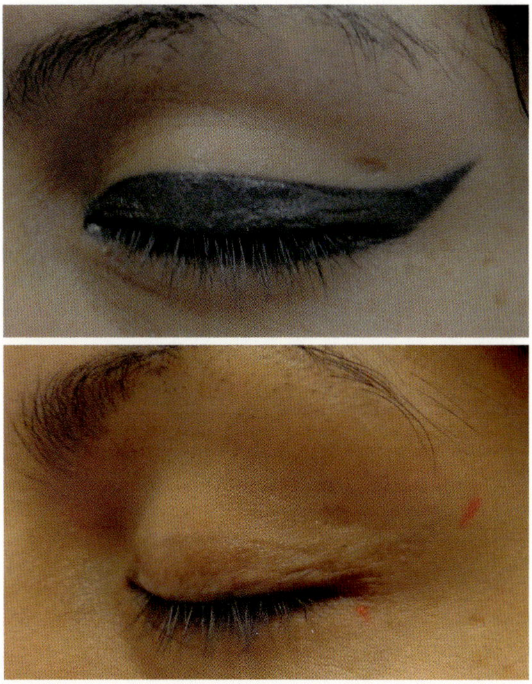

Fig. 9.4 Tattooed eyeliner. Before and after treatment with the QS 1064 nm Nd:YAG

transepidermal elimination of unwanted tattoo pigment. Another study found that the short-pulse erbium-doped yttrium aluminum garnet (SP Er:YAG) laser was superior to the Q-switched Nd:YAG laser and Q-switched alexandrite laser for removing cosmetic tattoos of white, flesh-colored, and brown inks [35]. With the Q-switched lasers, all three pigments darkened initially and then resolved gradually requiring up to 20, 18, and 10 sessions to remove white, flesh-colored, and brown tattoos, respectively. Only six sessions were required with the SP Er:YAG laser.

9.4 Vascular Concerns of the Periorbital Skin

Numerous vascular concerns of the periorbital skin are effectively treated with lasers, specifically the pulsed dye laser (PDL) or KTP laser, as their wavelengths effectively target hemoglobin. Common examples include superficial infantile hemangiomas, capillary vascular malformations, venous malformations, spider angiomas, cherry angiomas, telangiectasias, reticular veins, pyogenic granulomas, and purpura.

Hemangiomas are benign proliferative vascular tumors of endothelial tissue that affect 2–3 % of newborns and up to 10 % of infants within the first year [36]. The majority affect the head and neck, with 16 % of facial hemangiomas involving the eyelid [37]. They may present as superficial, deep, or compound (superficial and deep) and display many months of a proliferative phase followed by spontaneous involution at rates of about 10 % per year. Particular attention must be paid to deep and compound hemangiomas around the eye because of potential for amblyopia from anisometropia, strabismus, and obstruction, all of which can be exacerbated during the hemangioma's proliferative phase [38]. Despite involution, residual cosmetically undesirable effects are common in any form of hemangioma. Some report textural changes in up to 50 % of hemangiomas after involution [39]. Long-term residua from hemangiomas are more common when involution occurs over a longer period of time, and unfortunately no methods of identifying rate of involution presently exist [36]. The hemangioma and potential residua are recognized as causing psychological strain in children and family members. In general, laser therapy is not the ideal choice for deep hemangiomas because of their limited depth of penetration. However, for superficial hemangiomas, PDL is an excellent treatment option as it is safe and effective and minimizes extent of proliferation and residua if treated early.

A report of 22 patients highlights the value of early treatment of superficial eyelid hemangiomas with the 595 nm PDL [40]. These patients underwent 2–14 treatments, initiating therapy at 5–28 weeks of age. 77.3 % received an improvement rating of excellent (76–100 % improvement) and 36 % demonstrated complete clearance. No scarring, atrophy, hypopigmentation, infections, or ulcerations occurred during the study period, with the only side effect being hyperpigmentation in two subjects. Catalyzing its resolution and presumably limiting the proliferative phase likely contributed to the patients having no hemangioma residua. This report is in contrast to historical reports that resulted in side effects, particularly atrophy and hypopigmentation, but these side effects are attributable to the use of higher fluences, smaller spot sizes, absence of epidermal cooling, and different PDL wavelengths [41]. Two examples of the efficacy of treating superficial infantile hemangiomas with the PDL on the eyelid are shown in Figs. 9.5 and 9.6.

Vascular malformations are localized defects of vascular morphogenesis, which is in contrast to the neoplastic nature of hemangiomas. They are categorized by their anomalous vessels (e.g., capillary, venous, arterial, lymphatic) and by whether they have a fast (arterial) or slow flow.

Capillary vascular malformations (CVMs), often referred to as port-wine stains, are observed in 0.03 % of the population [42]. Facial CVMs classically course along the distribution of trigeminal nerve sensory branches, namely, V1 (ophthalmic), V2 (maxillary), and V3 (mandibular) branches. When present, especially around the eye, risks of coincident glaucoma and choroidal vascular malformations exist, as do concerns for syndromic capillary venous malformations

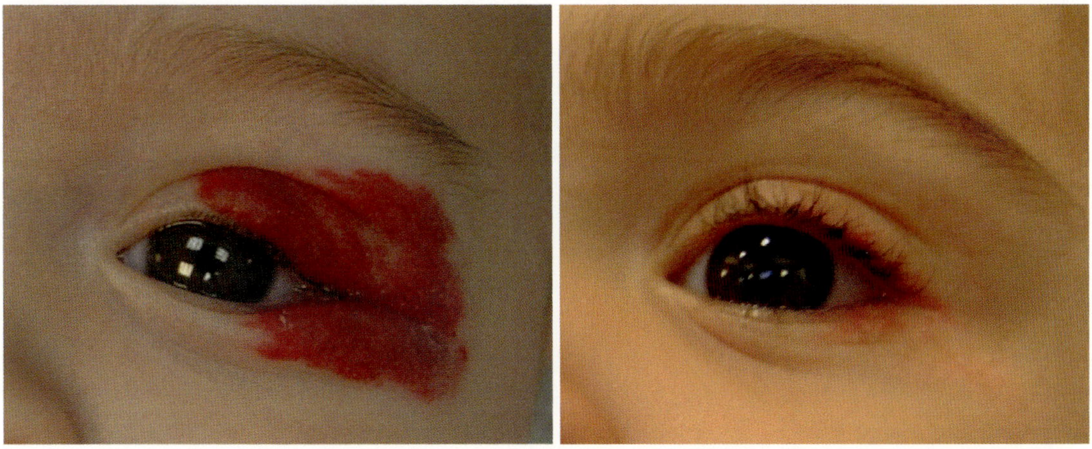

Fig. 9.5 Infantile hemangiomas of the periorbital area before and after treatment with the 595 nm pulsed dye laser

Fig. 9.6 Infantile hemangiomas of the periorbital area before and after treatment with the 595 nm pulsed dye laser

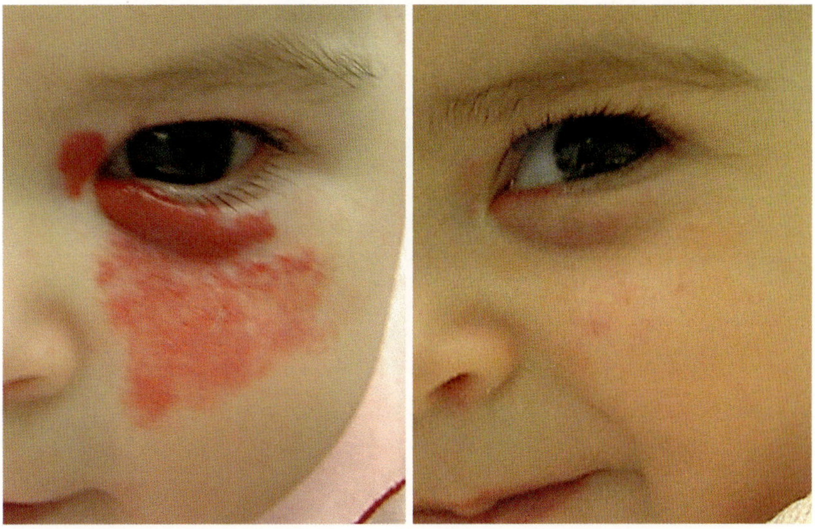

such as Sturge-Weber syndrome, von Hippel-Lindau syndrome, and Bonnet-Dechaume syndrome [43]. Over years and without treatment, CVMs typically develop vessel ectasia, which corresponds to the thickening, darkening, and cobblestoning appearance in aged lesions [44, 45]. Exuberant overgrowth can potentially lead to visual field obstruction of the eye or airway depending on location. The PDL is an important therapy in the treatment of periorbital CVMs and should be considered a treatment of choice for flat or mildly hypertrophic lesions [43, 46]. Other technologies are helpful, however, as intense pulsed light and the alexandrite lasers, among others, have demonstrated efficacy and even advantage in some situations [47, 48]. Early treatment has been shown to be safe and more effective [46]. Although anatomic differences do exist in terms of response to laser and light treatments, periorbital CVMs tend to respond well [49]. Efficacy from treatment with the 595 nm PDL can be appreciated in Fig. 9.7.

Venous malformations are examples of abnormal venous morphogenesis. While sclerotherapy with or without surgical excision is an important therapeutic option to consider, relatively long-wavelength lasers that penetrate more deeply into cutaneous veins may serve as an effective

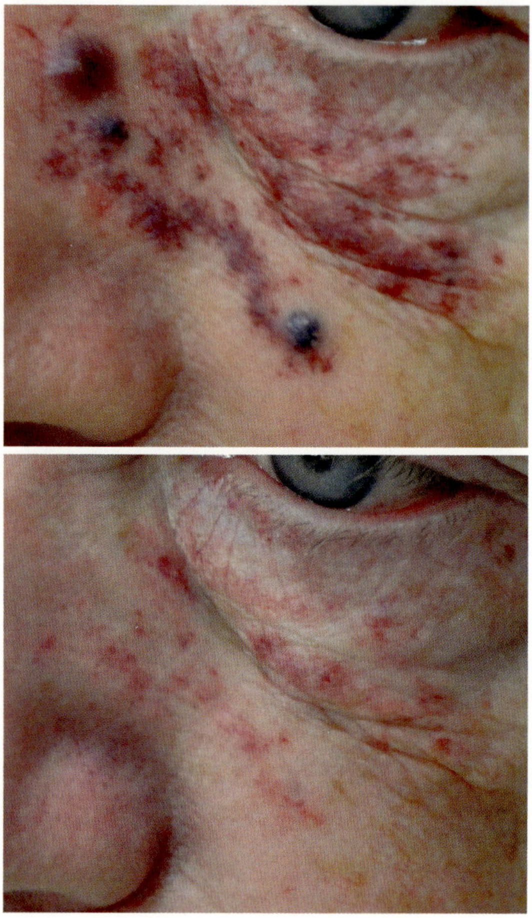

Fig. 9.7 Port-wine stain of the periorbital area before and after completed treatment with the 595 nm pulsed dye laser

therapeutic option as well [50]. When faced with appropriate candidates in our practice, the authors of this chapter often utilize a long-pulsed 532 or 1064 nm KTP laser with effective results.

Other collections of superficial vessels, including spider angiomas, cherry angiomas, telangiectasias, reticular veins, and pyogenic granulomas, also respond well to laser therapy [51–56].

9.5 Periorbital Purpura After Procedures

A relatively new use of laser is for the treatment of purpura, which is a common occurrence following periorbital procedures [57]. Most patients requiring medically necessary surgical interventions generally expect and accept purpura. However, purpura from a cosmetic procedure is often more frustrating for patients, as many of these patients demand little downtime or choose to have the procedure before important social events. Whatever the cause of purpura, the PDL can effectively accelerate the resolution. This was demonstrated in a study of ten adults with far more rapid resolution of purpura after treatment with a 595 nm PDL at spot size of 10 mm, fluence of 7.5 J/cm^2, and pulse duration of 6 ms.

9.6 Infraorbital Dark Circles

Infraorbital dark circles do not stem from a single pathology, but rather are multifactorial. The clinical appearance may result from hyperpigmentation, more translucent skin overlying vessels and muscles, or relative shadowing because of skin laxity, pronounced tear troughs, or pseudoherniation of infraorbital fat pads. To achieve an aesthetic improvement, the cause or causes must be identified in order to choose the correct treatment approach. Fortunately, laser technologies offer several options for the treatment of these concerns.

Placing traction on the infraorbital cheek assists in deciphering the cause. When applying downward traction on the infraorbital cheek, if the pigmented area grows proportionally with stretching without blanching or lightening in color, excessive pigmentation is likely [58, 59]. If traction causes spread of pigment, yet the area develops a deeper violaceous hue, then thin and translucent skin is likely the cause [60]. In other words, as the skin is stretched thin, the underlying vascular structures are more visible. Finally, if traction on the cheek diminishes the darkening, especially in a brightly lit room, shadowing is likely the driver.

Because of the diversity of causes, a variety of treatment modalities exist for dark circles. Surgery and filler have their role, as do topical bleaching agents, such as hydroquinone and retinoids. Fortunately, newer technologies exist which might contribute to therapy. Pigment lasers, such as the Q-switched ruby, alexandrite, and Nd:YAG, have demonstrated efficacy [58, 61, 62]. The 1064 nm Q-switched Nd:YAG shows particular value, as it appears to treat both melanocytic and vascular components of dark circles [60].

Transparency, laxity, fat pseudoherniation, and festooning may be improved with the resurfacing strategies for rejuvenation discussed earlier in this chapter. Resurfacing has also been shown to not only improve laxity and rhytides but also hyperpigmentation [63]. These tactics include ablative and non-ablative fractional resurfacing as well as radio frequency. With these treatments, dermal collagen remodeling may contribute to a thicker dermis and thereby diminish visibility of underlying vessels and musculature. Additionally, subsequent tightening minimizes shadowing.

9.7 Xanthelasma

Xanthelasma, or xanthelasma palpebrarum, is a soft, yellow papule and plaque involving the periorbital skin. Histologically, the lesions consist of foamy, lipid-laden histiocytes. For some, xanthelasma is a sign of hyperlipidemia and for a small minority a sign of familial hypercholesterolemia. While most patients with xanthelasma are normolipidemic, there is new evidence that normolipidemic patients with xanthelasma have similar cardiovascular risk to hyperlipidemic patients and should therefore be fully investigated in order to allow detection and early management of such risk [64]. Benefits of diet or medical therapy to treat xanthelasma are minimal, leaving most patients to rely on surgical or laser intervention if removal is desired [65]. Traditionally destructive methods, such as cryotherapy, chemical peeling, scalpel surgery, and electrosurgery, have not delivered sustained results and bring substantial risks for scar, dyspigmentation, ectropion, and eyelid asymmetry.

Interestingly, several types of lasers have shown efficacy against xanthelasma, including those traditionally used for pigment, blood vessels, and resurfacing [66–70]. Ablative lasers were studied initially. One study of 23 patients with a cumulative total of 52 xanthelasma lesions assessed efficacy of an ultrapulse carbon dioxide (CO_2) laser with a follow-up period of 10 months [70]. One treatment cleared all lesions, although three patients developed recurrence and dyspigmentation was found in 17 %. Importantly, no scarring was reported. Another study investigated the use of the erbium/YAG laser in 15 patients with 33 xanthelasma lesions [66]. With one treatment, the authors report complete clearance without dyspigmentation or scarring.

Recently, in a case series, 20 lesions were reported to be removed after a single ultrapulse CO_2 laser (10,600 nm) treatment with only two patients developing recurrence during the follow-up period of 9 months. Both these patients had been treated earlier by different modalities in the past. Side effects included only post-inflammatory hyperpigmentation in two patients [69]. A recent study of twenty patients compared the efficacy of ablative fractional CO_2 laser to super-pulsed CO_2 laser and found that downtime was significantly shorter for lesions treated by fractional CO_2 compared with those treated by super-pulsed CO_2 laser [71]. Patient satisfaction was also significantly higher for lesions treated by fractional CO_2 laser, especially flat plaques of xanthelasma that occupied large surface areas, compared with those treated by super-pulsed CO_2. For giant xanthelasma palpebrarum, twelve patients were treated with ultrapulsed CO_2 in three to four sessions at 15-day intervals with complete resolution and only one recurrence at 6 months [72].

Despite success with ablative lasers, non-ablative alternatives are desirable to minimize risks even further as well as to bypass the need for wounding, injection of local anesthetic, and downtime. Some claim benefits of the 1064 nm Nd:YAG [73, 74]. However, a controversial subsequent report of 37 patients with 76 lesions found both the 1064 and 532 nm Nd:YAG ineffective, even with more aggressive parameters [68, 74]. However, another non-ablative alternative, specifically the PDL, has shown greater promise. In a study of 20 patients with 38 lesions, patients underwent 5 treatments with the 585 nm

PDL at 2–3-week intervals [67]. About two-thirds demonstrated greater than 50 % improvement and one quarter demonstrated greater than 75 % improvement. A novel use of the non-ablative 1,450 nm diode laser for the treatment of xanthelasma was reported with 12 (75 %) of the 16 patients achieving moderate to marked improvement [75].

Additionally, there is evidence that supports non-ablative fractional resurfacing as a means to remedy xanthelasma. In a report, a 52-year-old woman with 4 years of xanthelasma was treated with the 1550 nm erbium-doped fractional laser [76]. After 7 treatments at 4–11-week intervals, the patient achieved near total improvement.

9.8 Adnexal Structures of Periorbital Skin

Periorbital adnexal structures, both normal and abnormal, can be removed with success using lasers. These structures include unwanted normal hair, trichiasis, syringomas, and hidrocystomas.

Laser hair removal has become a practical and often permanent means to remove unwanted hair [77, 78]. Removal generally relies on pigment-specific lasers that target melanin in the hair follicle, although intense pulsed light devices are sometimes used with good response [79]. Hair follicle elimination and destruction are observed clinically and histologically following laser treatment [77]. Because pigment serves as the target chromophore, blonde or white hairs are not as responsive to treatment.

Abnormal hair may also be targeted, so long as it is still pigmented. One report demonstrated efficacy of periorbital laser hair removal in ten patients with eyelid trichiasis after treatment with the ruby laser. At settings of 3 J and a 3.5 mm spot size, the ruby laser completely eliminated eyelid trichiasis in 6 patients after 1–3 sessions. Another three patients achieved a partial response. The tenth patient was lost to follow-up. Importantly, there were no reported complications and the procedure was well tolerated.

Abnormal tumors of the adnexa, most commonly syringomas and hidrocystomas, are also treatable with lasers. Syringomas and hidrocystomas are benign adnexal neoplasms that may be solitary, multiple, or eruptive lesions [80–82]. Essentially, effective treatment involves lesion destruction. This could be achieved using excision, electrodesiccation, or dermabrasion, among other destructive methods, but these come with risks for scarring and dyspigmentation. The benefit of laser resurfacing as a means of lesion destruction is the minimization of complications in tandem with efficacy.

Periorbital syringomas are a therapeutic challenge as one must take into account the number of lesions and skin type to determine which treatment is most appropriate for each patient. In one study using an ablative carbon dioxide laser, ten patients with multiple periorbital syringomas were treated at 5 W, 0.2 s scan time, and 3 mm spot size [83]. Two to four passes over 1–4 treatment sessions were performed resulting in elimination of syringomas in all patients over a median follow-up period of 16 months. Adverse effects included transient erythema lasting 6–12 weeks in all patients and hyperpigmentation in a patient with type IV skin that resolved over 8–12 weeks.

A trial of fractional ablative CO_2 laser was performed to treat 35 patients with periorbital syringomas with two sessions of fractional ablative CO_2 laser at 1-month intervals. Laser fluences were delivered in two or three passes over the lower eyelids, using a pulse energy of 100 mJ and a density of 100 spots/cm^2. Clinical improvement after 2 months of treatment showed the majority of patients having some mild to moderate improvement with only three patients having greater than 75 % clearance [84]. This could be attributed to the nature of fractional devices when treating these lesions. When employing fractional devices, the pinpoint injury is not wide enough to cause complete lesional destruction. Therefore, multiple passes and consecutive treatments are needed to effectively reduce lesions [85]. Erbium laser ablation has shown efficacy against syringomas. In a study of 104 patients with a variety of skin lesions, some with syringomas, the erbium/YAG system successfully eliminated the lesions using a 0.350 ms pulse duration and 0.1–1.7 J [86]. The syringomas were successfully vaporized with minimal peripheral thermal damage and good to excellent cosmetic outcome.

Other more inventive methods have been published attempting to treat these periorbital syringomas. One study employed a multiple-drilling method using CO_2 laser for 11 patients with syringomas [87]. Rather than resurfacing or cutting the skin, the clinicians created several relatively deep holes with the ablative laser into the targeted lesions. This strategy was taken in an attempt to reach the deep components of the adnexal structures. Eleven patients were treated, 10 with periorbital lesions and one with vulvar lesions, over one to four treatment sessions. All patients were found to achieve good to excellent clinical responses. No serious complications were noted. Another group cleverly integrated temporary tattooing into the treatment of patients' syringomas [88]. In this report, multiple periorbital syringomas had their surface epithelium removed with a carbon dioxide laser. Afterward, droplets of black ink were laid on the syringomas and iontophoresis was performed to create a tattoo in the lesions. Finally, the Q-switched alexandrite laser was applied to the lesion with complete disappearance of the syringomas at the 1-week follow-up evaluation. The only significant adverse effect was hyperpigmentation lasting more than 2 months in a patient with type V skin.

Hidrocystomas have also been treated effectively with lasers. Surprisingly, some report successful treatment of hidrocystomas using PDL. This is unexpected since the PDL's target chromophore in a hidrocystoma is not known. In one report, a 585 nm pulsed dye laser was used at fluences of 7.0–7.5 J/cm^2 over 6- to 8-week intervals [89]. After 4 treatments, there was near total resolution of the lesions. Other reports, however, have not had such success with the PDL, raising questions about the real effectiveness of this strategy [90]. As would be predicted, however, hidrocystomas can be successfully treated using ablative lasers, such as the carbon dioxide laser [91]. Conceptually, this makes sense, as destruction of the cyst wall itself could lead to resolution of the lesion.

Despite success with lasers for these adnexal lesions, some groups still rely and endorse electrosurgery and excision [92, 93]. Certainly, these options are successful in some circumstances, such as cases of giant histiocytomas, and are relatively more accessible to clinicians. However, the authors of this chapter encourage clinicians not to choose these alternatives simply because laser systems may not be available in their immediate practice.

9.9 Safety and Complications (Including Contraindications)

The use of lasers around the eyes raises a number of serious safety concerns. Physicians must be fluent in these concerns and know the appropriate measures to protect themselves, their staff, and the patients. Ocular damage from inadvertent laser exposure is always a risk with lasers. Ocular melanin and vasculature are at particular risk when using lasers that target those chromophores. The cornea and sclera are at particular risk when using resurfacing lasers because of the high water content of these structures.

In practice, if any reasonable risk exists to the eyes, everyone must have protective eyewear. For the physician and staff, wraparound goggles should be worn that are rated as having an optical density (OD) of 4 or greater. OD is calculated as log (1/T) where T is the transmittance of light through the eyewear. The particular OD for each pair of goggles differs based on wavelength and should be specified directly on the glasses. One should not rely on the color of the protective goggles alone as a determinant of which pair to wear.

For patients, external or internal eye shields may be used. When the laser is not in or directed at the immediate eyelid area, external opaque shields should be adequate. Otherwise, internal shields are required. When choosing internal eye shields, non-reflective metal shields should be used. Internal plastic shields used by some surgeons during non-laser procedures do not adequately protect against most lasers, such as the carbon dioxide laser, since they may penetrate the shield. Pretreatment with ophthalmic anesthetic drops may alleviate patient discomfort, and internal eye shields are generally well tolerated.

Despite available safety protocols, complications from periorbital laser use are reported, especially when the appropriate precautions are not met [8, 94–100]. Complications include iris atrophy, posterior synechiae, iris pigment dispersion, anterior uveitis, ectropion, and blindness. In most reports of these cases, the patient simply closed their eyes, covered their eyes with their own fingers, or inadequately covered the eyes with displaced external shields. Several of the reports stem from laser hair removal of the lower aspect of the eyebrow. Often external eye shields were displaced or removed to allow a bulky laser tip to treat the target area. The laser's proximity in combination with Bell's phenomenon puts the patient's eyes at substantial risk when lasing the lower eyebrow.

Despite proper shielding, patients may still appreciate a flash of light concurrent with each periorbital laser pulse. The pulse is thought to somehow trigger the retinal photoreceptors. Safety concerns have been raised but the evidence does not show any harmful effects. In one study, five patients undergoing diode laser hair removal for severe trichiasis were evaluated with pre- and posttreatment ophthalmic exams [101]. These exams included slit-lamp, pupillary, funduscopic, and objective retinal electroretinogram studies. Although 3 of the 5 patients experienced the sensation of flashing lights during treatment, there was no detectable change in any of the listed exams after treatment.

Aside from ophthalmic risks, lasers carry intrinsic safety concerns for fires and burns, particularly when flammable materials such as paper drapes, alcohol, or supplied oxygen are used [102, 103]. Therefore, flammable materials should be removed from the treatment area. Additionally, when a sedated patient requires concentrated oxygen and/or nitrous oxygen, use of a laryngeal mask or endotracheal intubation limits release of the flammable gas. Moist surgical drapes contribute to fire hazard safety and may even be wrapped around the portion of a laryngeal mask or endotracheal tube exiting the mouth [103]. Aerosolization of infectious agents, like viruses, and tissue particles are also concerns with laser treatment. These risks are still being clarified [104], but appropriate ventilation, consistent vacuum use, gloves, and masks may assist in preventing consequences from these risks.

Conclusion

Lasers are invaluable for medical and cosmetic concerns around the eyes. Rejuvenation and the elimination of pigment, vascular lesions, dark circles, xanthelasma, and adnexal tumors are all possible with the appropriate use of lasers. With advances of existing technologies and the development of newer technologies on the horizon, periorbital concerns will continue to be more effectively and safely treated.

References

1. Maiman TH (1960) Stimulated optical radiation in Ruby. Nature 187(4736):493–494. doi:10.1038/187493a0
2. Anderson R, Parrish J (1983) Selective photothermolysis: precise microsurgery by selective absorption of pulsed radiation. Science 220(4596):524–527. doi:10.1126/science.6836297
3. Weiss ET, Brauer JA, Anolik R, Reddy KK, Karen JK, Hale EK, Brightman LA, Bernstein L, Geronemus RG (2013) 1927-nm fractional resurfacing of facial actinic keratoses: a promising new therapeutic option. J Am Acad Dermatol 68(1):98–102. doi:10.1016/j.jaad.2012.05.033
4. Brightman LA, Brauer JA, Anolik R, Weiss E, Karen J, Chapas A, Hale E, Bernstein L, Geronemus RG (2009) Ablative and fractional ablative lasers. Dermatol Clin 27(4):479–489. doi:10.1016/j.det.2009.08.009, vi–vii
5. Manstein D, Herron GS, Sink RK, Tanner H, Anderson RR (2004) Fractional photothermolysis: a new concept for cutaneous remodeling using microscopic patterns of thermal injury. Lasers Surg Med 34(5):426–438. doi:10.1002/lsm.20048
6. Sukal SA, Chapas AM, Bernstein LJ, Hale EK, Kim KH, Geronemus RG (2008) Eyelid tightening and improved eyelid aperture through nonablative fractional resurfacing. Dermatol Surg: Off Publi Am Soc Dermatol Surg [et al] 34(11):1454–1458. doi:10.1111/j.1524-4725.2008.34308.x
7. Kotlus BS (2010) Dual-depth fractional carbon dioxide laser resurfacing for periocular rhytidosis. Dermatol Surg: Off Publi Am Soc Dermatol Surg [et al] 36(5):623–628.doi:10.1111/j.1524-4725.2010.01516.x
8. Fife DJ, Fitzpatrick RE, Zachary CB (2009) Complications of fractional CO2 laser resurfacing:

four cases. Lasers Surg Med 41(3):179–184. doi:10.1002/lsm.20753
9. Behroozan DS, Goldberg LH, Dai T, Geronemus RG, Friedman PM (2006) Fractional photothermolysis for the treatment of surgical scars: a case report. J Cosmetic Laser Therapy: Off Publ Eur Soc Laser Dermatol 8(1):35–38. doi:10.1080/14764170600607251
10. Kunishige JH, Katz TM, Goldberg LH, Friedman PM (2010) Fractional photothermolysis for the treatment of surgical scars. Dermatol Surg: Off Publi Am Soc Dermatol Surg [et al] 36(4):538–541. doi:10.1111/j.1524-4725.2010.01491.x
11. Weiss ET, Chapas A, Brightman L, Hunzeker C, Hale EK, Karen JK, Bernstein L, Geronemus RG (2010) Successful treatment of atrophic postoperative and traumatic scarring with carbon dioxide ablative fractional resurfacing: quantitative volumetric scar improvement. Arch Dermatol 146(2):133–140. doi:10.1001/archdermatol.2009.358
12. Bastiaens M, ter Huurne J, Gruis N, Bergman W, Westendorp R, Vermeer BJ, Bouwes Bavinck JN (2001) The melanocortin-1-receptor gene is the major freckle gene. Hum Mol Genet 10(16):1701–1708
13. Bastiaens M, Hoefnagel J, Westendorp R, Vermeer BJ, Bouwes Bavinck JN (2004) Solar lentigines are strongly related to sun exposure in contrast to ephelides. Pigment Cell Res/sponsored by the European Society for Pigment Cell Research and the International Pigment Cell Society 17(3):225–229. doi:10.1111/j.1600-0749.2004.00131.x
14. Lodish MB, Stratakis CA (2011) The differential diagnosis of familial lentiginosis syndromes. Fam Cancer 10(3):481–490. doi:10.1007/s10689-011-9446-x
15. Taylor CR, Anderson RR (1993) Treatment of benign pigmented epidermal lesions by Q-switched Ruby laser. Int J Dermatol 32(12):908–912. doi:10.1111/j.1365-4362.1993.tb01417.x
16. Shimbashi T, Kamide R, Hashimoto T (1997) Long-term follow-up in treatment of Solar Lentigo and Café-au-Lait Macules with Q-switched Ruby laser. Aesthetic Plast Surg 21(6):445–448. doi:10.1007/s002669900155
17. Brauer JA, McDaniel DH, Bloom BS, Reddy KK, Bernstein LJ, Geronemus RG (2014) Nonablative 1927 nm fractional resurfacing for the treatment of facial photopigmentation. J Drugs Dermatol 13(11):1317–1322
18. Price HN, Schaffer JV (2010) Congenital melanocytic nevi-when to worry and how to treat: facts and controversies. Clin Dermatol 28(3):293–302. doi:10.1016/j.clindermatol.2010.04.004
19. Soden CE, Smith K, Skelton H (2001) Histologic features seen in changing nevi after therapy with an 810 nm pulsed diode laser for hair removal in patients with dysplastic nevi. Int J Dermatol 40(8):500–504. doi:10.1046/j.1365-4362.2001.01251.x
20. Lim JY, Jeong Y, Whang KK (2009) A combination of dual-mode 2,940 nm Er:YAG laser ablation with surgical excision for treating medium-sized congenital melanocytic nevus. Ann Dermatol 21(2):120. doi:10.5021/ad.2009.21.2.120
21. Rajpar SF, Abdullah A, Lanigan SW (2007) Er:YAG laser resurfacing for inoperable medium-sized facial congenital melanocytic naevi in children. Clin Exp Dermatol 32(2):159–161. doi:10.1111/j.1365-2230.2006.02286.x
22. Kishi K, Okabe K, Ninomiya R, Konno E, Hattori N, Katsube K, Imanish N, Nakajima H, Nakajima T (2009) Early serial Q-switched ruby laser therapy for medium-sized to giant congenital melanocytic naevi. Br J Dermatol 161(2):345–352. doi:10.1111/j.1365-2133.2009.09153.x
23. Lee SE, Choi JY, Hong KT, Lee KR (2015) Treatment of acquired and small congenital melanocytic nevi with combined Er: YAG laser and long-pulsed alexandrite laser in Asian skin. Dermatol Surg: Off Publi Am Soc Dermatol Surg [et al] 41(4):473–480. doi:10.1097/dss.0000000000000288
24. Franceschini D, Dinulos JG (2015) Dermal melanocytosis and associated disorders. Curr Opin Pediatr 26(4):480–485. doi:10.1097/mop.0000000000000247
25. Magarasevic L, Abazi Z (2013) Unilateral open-angle glaucoma associated with the ipsilateral nevus of ota. Case Rep Ophthalmol Med 2013:924937. doi:10.1155/2013/924937
26. Teekhasaenee C (1990) Ocular findings in oculodermal melanocytosis. Arch Ophthalmol 108(8):1114. doi:10.1001/archopht.1990.01070100070037
27. Chan HH, Ying S-Y, Ho W-S, Kono T, King WW (2000) An in vivo trial comparing the clinical efficacy and complications of Q-switched 755 nm alexandrite and Q-switched 1064 nm Nd:YAG lasers in the treatment of nevus of ota. Dermatol Surg 26(10):919–922. doi:10.1046/j.1524-4725.2000.026010919.x
28. Lowe NJ, Wieder JM, Sawcer D, Burrows P, Chalet M (1993) Nevus of Ota: treatment with high energy fluences of the Q-switched ruby laser. J Am Acad Dermatol 29(6):997–1001. doi:10.1016/0190-9622(93)70280-7
29. Tse Y, Levine VJ, McClain SA, Ashinoff R (1994) The removal of cutaneous pigmented lesions with the Q-switched ruby laser and the Q-switched neodymium: yttrium-aluminum-garnet laser. J Dermatol Surg Oncol20(12):795–800.doi:10.1111/j.1524-4725.1994.tb03707.x
30. Wang H-W, Liu Y-H, Zhang G-K, Jin H-Z, Zuo Y-G, Jiang G-T, Wang J-B (2007) Analysis of 602 Chinese cases of nevus of Ota and the treatment results treated by Q-switched alexandrite laser. Dermatol Surg 33(4):455–460.doi:10.1111/j.1524-4725.2007.33093.x
31. Chan HH, L-k L, Wong DSY, Leung RSC, S-y Y, C-f L, W-s H, Chua JKH (2001) Nevus of Ota: a new classification based on the response to laser treatment. Lasers Surg Med 28(3):267–272. doi:10.1002/lsm.1049
32. Anderson RR (1993) Cosmetic tattoo ink darkening. Arch Dermatol 129(8):1010. doi:10.1001/archderm.1993.01680290082012

33. Geronemus RG (1996) Surgical pearl: Q-switched Nd:YAG laser removal of eyeliner tattoo. J Am Acad Dermatol 35(1):101–102. doi:10.1016/s0190-9622(96)90504-6
34. Mafong EA, Kauvar AN, Geronemus RG (2003) Surgical pearl: removal of cosmetic lip-liner tattoo with the pulsed carbon dioxide laser. J Am Acad Dermatol 48(2):271–272. doi:10.1067/mjd.2003.29
35. Wang CC, Huang CL, Yang AH, Chen CK, Lee SC, Leu FJ (2010) Comparison of two Q-switched lasers and a short-pulse erbium-doped yttrium aluminum garnet laser for treatment of cosmetic tattoos containing titanium and iron in an animal model. Dermatol Surg: Off Publi Am Soc Dermatol Surg [et al] 36(11):1656–1663. doi:10.1111/j.1524-4725.2010.01714.x
36. Collins Finn M, Glowacki J, Mulliken JB (1983) Congenital vascular lesions: clinical application of a new classification. J Pediatr Surg 18(6):894–900. doi:10.1016/s0022-3468(83)80043-8
37. Waner M, North PE, Scherer KA, Frieden IJ, Waner A, Mihm MC (2003) The nonrandom distribution of facial hemangiomas. Arch Dermatol 139(7):869–875. doi:10.1001/archderm.139.7.869
38. Ceisler E, Blei F (2003) Ophthalmic issues in hemangiomas of infancy. Lymphat Res Biol 1(4):321–330. doi:10.1089/153968503322758148
39. Mulliken JB, Fishman SJ, Burrows PE (2000) Vascular anomalies. Curr Probl Surg 37(8):517–584. doi:10.1016/s0011-3840(00)80013-1
40. Hunzeker CM, Geronemus RG (2010) Treatment of superficial infantile hemangiomas of the eyelid using the 595-nm pulsed dye laser. Dermatol Surg: Off Publi Am Soc Dermatol Surg [et al] 36(5):590–597. doi:10.1111/j.1524-4725.2010.01511.x
41. Batta K, Goodyear HM, Moss C, Williams HC, Hiller L, Waters R (2002) Randomised controlled study of early pulsed dye laser treatment of uncomplicated childhood haemangiomas: results of a 1-year analysis. Lancet 360(9332):521–527. doi:10.1016/s0140-6736(02)09741-6
42. Jacobs AH, Walton RG (1977) The incidence of birthmarks in the Neonate. Obstet Gynecol Surv 32(2):94–95. doi:10.1097/00006254-197702000-00014
43. Holy A (1992) Treatment of periorbital port-wine stains with the flashlamp-pumped pulsed dye laser. Arch Ophthalmol 110(6):793. doi:10.1001/archopht.1992.01080180065029
44. Barsky SH, Rosen S, Geer DE, Noe JM (1980) The nature and evolution of port wine stains: a computer-assisted study. J Invest Dermatol 74(3):154–157. doi:10.1111/1523-1747.ep12535052
45. Geronemus RG, Ashinoff R (1991) The medical necessity of evaluation and treatment of port-wine stains. J Dermatol Surg Oncol 17(1):76–79
46. Chapas AM, Eickhorst K, Geronemus RG (2007) Efficacy of early treatment of facial port wine stains in newborns: a review of 49 cases. Lasers Surg Med 39(7):563–568. doi:10.1002/lsm.20529
47. Faurschou A, Togsverd-Bo K, Zachariae C, Haedersdal M (2009) Pulsed dye laser vs. intense pulsed light for port-wine stains: a randomized side-by-side trial with blinded response evaluation. Br J Dermatol 160(2):359–364. doi:10.1111/j.1365-2133.2008.08993.x
48. Izikson L, Nelson JS, Anderson RR (2009) Treatment of hypertrophic and resistant port wine stains with a 755 nm laser: a case series of 20 patients. Lasers Surg Med 41(6):427–432. doi:10.1002/lsm.20793
49. Renfro L (1993) Anatomical differences of port-wine stains in response to treatment with the pulsed dye laser. Arch Dermatol 129(2):182. doi:10.1001/archderm.1993.01680230066007
50. Bagazgoitia L, Boixeda P, Lopez-Caballero C, Beà S, Santiago JL, Jaén P (2008) Venous malformation of the eyelid treated with pulsed-dye-1064-nm neodymium yttrium aluminum garnet sequential laser: an effective and safe treatment. Ophthal Plas Reconst Surg 24(6):488–490. doi:10.1097/iop.0b013e31818bed57
51. Hare McCoppin HH, Goldberg DJ (2010) Laser treatment of facial telangiectases: an update. Dermatol Surg 36(8):1221–1230. doi:10.1111/j.1524-4725.2010.01613.x
52. Dawn G, Gupta G (2003) Comparison of potassium titanyl phosphate vascular laser and hyfrecator in the treatment of vascular spiders and cherry angiomas. Clin Exp Dermatol 28(6):581–583. doi:10.1046/j.1365-2230.2003.01352.x
53. Collyer J, Boone SL, White LE, Rademaker A, West DP, Anderson K, Kim NA, Smith S, Yoo S, Alam M (2010) Comparison of treatment of cherry angiomata with pulsed-dye laser, potassium titanyl phosphate laser, and electrodesiccation: a randomized controlled trial. Arch Dermatol 146(1):33–37. doi:10.1001/archdermatol.2009.318
54. Lai SW, Goldman MP (2007) Treatment of facial reticular veins with dynamically cooled, variable spot-sized 1064 nm Nd:YAG laser. J Cosmet Dermatol 6(1):6–8. doi:10.1111/j.1473-2165.2007.00256.x
55. Clark C, Cameron H, Moseley H, Ferguson J, Ibbotson SH (2004) Treatment of superficial cutaneous vascular lesions: experience with the KTP 532 nm laser. Lasers Med Sci 19(1):1–5. doi:10.1007/s10103-004-0294-x
56. Sud AR, Tan ST (2010) Pyogenic granuloma–treatment by shave-excision and/or pulsed-dye laser. J Plast Reconstr Aesthet Surg 63(8):1364–1368. doi:10.1016/j.bjps.2009.06.031
57. Karen JK, Hale EK, Geronemus RG (2010) A simple solution to the common problem of ecchymosis. Arch Dermatol 146(1):94–95. doi:10.1001/archdermatol.2009.343
58. Friedmann DP, Goldman MP (2015) Dark circles: etiology and management options. Clin Plast Surg 42(1):33–50. doi:10.1016/j.cps.2014.08.007
59. Epstein JS (1999) Management of infraorbital dark circles: a significant cosmetic concern. Arch Facial Plast Surg 1(4):303–307. doi:10.1001/archfaci.1.4.303
60. Roh MR, Chung KY (2009) Infraorbital dark circles: definition, causes, and treatment options. Dermatol Surg 35(8):1163–1171. doi:10.1111/j.1524-4725.2009.01213.x

61. Momosawa A, Kurita M, Ozaki M, Miyamoto S, Kobayashi Y, Ban I, Harii K (2008) Combined therapy using Q-switched ruby laser and bleaching treatment with tretinoin and hydroquinone for periorbital skin hyperpigmentation in Asians. Plast Reconstr Surg 121(1):282–288. doi:10.1097/01.prs.0000293869.00522.ec
62. Watanabe S, Nakai K, Ohnishi T (2006) Condition known as "dark rings under the eyes" in the Japanese population is a kind of dermal melanocytosis which can be successfully treated by Q-switched ruby laser. Dermatol Surg 32(6):785–789. doi:10.1111/j.1524-4725.2006.32161.x
63. West TB, Alster TS (1998) Improvement of infraorbital hyperpigmentation following carbon dioxide laser resurfacing. Dermatol Surg 24(6):615–616. doi:10.1111/j.1524-4725.1998.tb04216.x
64. Esmat S, Abdel-Halim MR, Fawzy MM, Nassef S, Esmat S, Ramzy T, El Fouly ES (2015) Are normolipidaemic patients with xanthelasma prone to atherosclerosis? Clin Exp Dermatol 40(4):373–378. doi:10.1111/ced.12594
65. Rohrich RJ, Janis JE, Pownell PH (2002) Xanthelasma palpebrarum: a review and current management principles. Plast Reconstr Surg 110(5):1310–1314. doi:10.1097/00006534-200210000-00016
66. Borelli C, Kaudewitz P (2001) Xanthelasma palpebrarum: treatment with the erbium:YAG laser. Lasers Surg Med 29(3):260–264. doi:10.1002/lsm.1117
67. Karsai S, Czarnecka A, Raulin C (2010) Treatment of xanthelasma palpebrarum using a pulsed dye laser. Dermatol Surg 36(5):610–617. doi:10.1111/j.1524-4725.2010.01514.x
68. Karsai S, Schmitt L, Raulin C (2009) Is Q-switched neodymium-doped yttrium aluminium garnet laser an effective approach to treat xanthelasma palpebrarum? Results from a clinical study of 76 cases. Dermatol Surg 35(12):1962–1969. doi:10.1111/j.1524-4725.2009.01314.x
69. Pathania V, Chatterjee M (2015) Ultrapulse carbon dioxide laser ablation of xanthelasma palpebrarum: a case series. J Cutan Aesthetic Surg 8(1):46–49. doi:10.4103/0974-2077.155084
70. Raulin C, Schoenermark MP, Werner S, Greve B (1999) Xanthelasma palpebrarum: treatment with the ultrapulsed CO2 laser. Lasers Surg Med 24(2):122–127. doi:10.1002/(sici)1096-9101(1999)24:2<122::aid-lsm7>3.0.co;2-6
71. Esmat SM, Elramly AZ, Abdel Halim DM, Gawdat HI, Taha HI (2014) Fractional CO2 laser is an effective therapeutic modality for xanthelasma palpebrarum: a randomized clinical trial. Dermatol Surg: Off Publi Am Soc Dermatol Surg [et al] 40(12):1349–1355. doi:10.1097/dss.0000000000000172
72. Corradino B, Di Lorenzo S, Triolo A, Moschella F (2014) Laser treatment of giant xanthelasma palpebrarum. Lasers Med Sci. doi:10.1007/s10103-014-1664-7
73. Fusade T (2007) Treatment of xanthelasma palpebrarum by 1064-nm Q-switched Nd:YAG laser: a studyof11cases.BrJDermatol0(0):071004160508022-???. doi:10.1111/j.1365-2133.2007.08194.x
74. Fusade T (2011) About the treatment of xanthelasma palpebrarum using a 1,064 Q-switched neodymium-doped yttrium aluminum garnet laser. Dermatol Surg: Off Publi Am Soc Dermatol Surg [et al] 37(3):403–404. doi:10.1111/j.1524-4725.2011.01899.x
75. Park EJ, Youn SH, Cho EB, Lee GS, Hann SK, Kim KH, Kim KJ (2011) Xanthelasma palpebrarum treatment with a 1,450-nm-diode laser. Dermatol Surg: Off Publi Am Soc Dermatol Surg [et al] 37(6):791–796. doi:10.1111/j.1524-4725.2011.01945.x
76. Katz TM (2009) Fractional photothermolysis. Arch Dermatol 145(10):1091. doi:10.1001/archdermatol.2009.234
77. Grossman MC, Dierickx C, Farinelli W, Flotte T, Anderson RR (1996) Damage to hair follicles by normal-mode ruby laser pulses. J Am Acad Dermatol 35(6):889–894. doi:10.1016/s0190-9622(96)90111-5
78. Ibrahimi OA, Avram MM, Hanke CW, Kilmer SL, Anderson RR (2011) Laser hair removal. Dermatol Ther 24(1):94–107. doi:10.1111/j.1529-8019.2010.01382.x
79. Toosi P, Sadighha A, Sharifian A, Razavi GM (2006) A comparison study of the efficacy and side effects of different light sources in hair removal. Lasers Med Sci 21(1):1–4. doi:10.1007/s10103-006-0373-2
80. D'Andrea M, Reggiani C, Fasano D, Betts CM, Montanari F, Lanzoni A, Reggiani M, Foschini MP (2013) Tumours of the skin adnexa: a case series with focus on multiple segmental forms. Pathologica 105(6):337–341
81. Patrizi INSMEA (1998) Syringoma: a review of twenty-nine cases. Acta Derm Venereol 78(6):460–462. doi:10.1080/000155598442791
82. Smith JD (1973) Hidrocystomas. Arch Dermatol 108(5):676. doi:10.1001/archderm.1973.01620260026008
83. Wang JI, Roenigk HH (1999) Treatment of multiple facial syringomas with the carbon dioxide (CO2) laser. Dermatol Surg 25(2):136–139. doi:10.1046/j.1524-4725.1999.08111.x
84. Cho SB, Kim HJ, Noh S, Lee SJ, Kim YK, Lee JH (2011) Treatment of syringoma using an ablative 10,600-nm carbon dioxide fractional laser: a prospective analysis of 35 patients. Dermatol Surg: Off Publi Am Soc Dermatol Surg [et al] 37(4):433–438. doi:10.1111/j.1524-4725.2011.01915.x
85. Brightman L, Geronemus R (2011) Commentary: treatment of syringoma using an ablative 10,600-nm carbon dioxide fractional laser. Dermatol Surg: Off Publi Am Soc Dermatol Surg [et al] 37(4):439–440. doi:10.1111/j.1524-4725.2011.01936.x
86. Riedel F, Windberger J, Stein E, Hormann K (1998) Treatment of periocular skin changes with the erbium: YAG laser. Der Ophthalmologe 95(11):771–775. doi:10.1007/s003470050351
87. Park HJ, Lee D-Y, Lee J-H, Yang J-M, Lee ES, Kim W-S (2007) The treatment of syringomas by CO 2 laser using a multiple-drilling method. Dermatol Surg 33(3):310–313.doi:10.1111/j.1524-4725.2007.33065.x
88. Park HJ, Lim SH, Kang HA, Byun DG, Houh D (2001) Temporary tattooing followed by Q-switched

alexandrite laser for treatment of syringomas. Dermatol Surg 27(1):28–30. doi:10.1046/j.1524-4725.2001.00188.x
89. Tanzi E, Alster TS (2001) Pulsed dye laser treatment of multiple eccrine hidrocystomas: a novel approach. Dermatol Surg 27(10):898–900. doi:10.1046/j.1524-4725.2001.01078.x
90. Choi JE, Ko NY, Son SW (2007) Lack of effect of the pulsed-dye laser in the treatment of multiple eccrine hidrocystomas. Dermatol Surg 33(12):1513–1515. doi:10.1097/00042728-200712000-00018
91. Madan V, August PJ, Ferguson J (2009) Multiple eccrine hidrocystomas--response to treatment with carbon dioxide and pulsed dye lasers. Dermatol Surg: Off Publi Am Soc Dermatol Surg [et al] 35(6):1015–1017. doi:10.1111/j.1524-4725.2009.01177.x
92. Al Aradi IK (2006) Periorbital syringoma: a pilot study of the efficacy of low-voltage electrocoagulation. Dermatol Surg 32(10):1244–1250. doi:10.1111/j.1524-4725.2006.32284.x
93. Gupta S, Handa U, Handa S, Mohan H (2001) The efficacy of electrosurgery and excision in treating patients with multiple apocrine hidrocystomas. Dermatol Surg 27(4):382–384. doi:10.1046/j.1524-4725.2001.00210.x
94. Lin CC, Tseng PC, Chen CC, Woung LC, Liou SW (2011) Iritis and pupillary distortion after periorbital cosmetic alexandrite laser. Graefe's archive for clinical and experimental ophthalmology=Albrecht von Graefes Archiv fur klinische und experimentelle Ophthalmologie 249(5):783–785. doi:10.1007/s00417-010-1554-z
95. Lee WW, Murdock J, Albini TA, O'Brien TP, Levine ML (2011) Ocular damage secondary to intense pulse light therapy to the face. Ophthal Plast Reconstr Surg 27(4):263–265. doi:10.1097/IOP.0b013e31820c6e23
96. Elkin Z, Ranka MP, Kim ET, Kahanowicz R, Whitmore WG (2011) Iritis and iris atrophy after eyebrow epilation with alexandrite laser. Clinical Ophthalmol (Auckland, NZ) 5:1733–1735. doi:10.2147/opth.s26035
97. Halkiadakis I, Skouriotis S, Stefanaki C, Patsea E, Papakonstatndinou D, Amariotakis A, Georgopoulos GT (2007) Iris atrophy and posterior synechiae as a complication of eyebrow laser epilation. J Am Acad Dermatol 57(2):S4–S5. doi:10.1016/j.jaad.2006.07.024
98. Hammes S, Augustin A, Raulin C, Ockenfels H-M, Fischer E (2007) Pupil damage after periorbital laser treatment of a port-wine stain. Arch Dermatol 143(3):392–394. doi:10.1001/archderm.143.3.392
99. Shulman S, Bichler I (2009) Ocular complications of laser-assisted eyebrow epilation. Eye 23(4):982–983. doi:10.1038/eye.2008.436
100. Le Jeune M, Autie M, Monnet D, Brezin AP (2007) Ocular complications after laser epilation of eyebrows. Eur J Dermatol 17(6):553–554. doi:10.1684/ejd.2007.0287
101. Pham RTH, Tzekov RT, Biesman BS, Marmor MF (2002) Retinal evaluation after 810 nm dioderm laser removal of eyelashes. Dermatol Surg 28(9):836–840. doi:10.1046/j.1524-4725.2002.02032.x
102. Yardley IE, Donaldson LJ (2010) Surgical fires, a clear and present danger. Surgeon: J Royal Coll Surg Edinburgh Ireland 8(2):87–92. doi:10.1016/j.surge.2010.01.005
103. Waldorf HA, Kauvar NB, Geronemus RG, Leffel DJ (1996) Remote fire with the pulsed dye laser: risk and prevention. J Am Acad Dermatol 34(3):503–506
104. Weyandt GH, Tollmann F, Kristen P, Weissbrich B (2011) Low risk of contamination with human papilloma virus during treatment of condylomata acuminata with multilayer argon plasma coagulation and CO(2) laser ablation. Arch Dermatol Res 303(2):141–144. doi:10.1007/s00403-010-1119-3

10 PRP for Lip and Eye Rejuvenation

Gabriella Fabbrocini, Maria Carmela Annunziata, Caterina Mazzella, and Saverio Misso

10.1 Introduction

Platelet-rich plasma (PRP) is an autologous concentration of human platelets contained in a small volume of plasma. Platelets can be likened to cell reservoirs that produce, store, and, finally, release numerous growth factors capable of stimulating the proliferation of stem cells and the replication of mesenchymal cells, fibroblasts, osteoblasts, and endothelial cells.

PRP is composed of several different growth factors, platelet-derived growth factor (PDGF), transforming growth factor a (TGF-a), vascular endothelial growth factor (VEGF), insulin-like growth factor 1 (IGF-1), epidermal growth factor (EGF), basic fibroblast growth factor (bFGF), transforming growth factor b1 (TGF-b1), and platelet-activating factor (PAF), that are released through degranulation and stimulate bone and soft tissue healing. The secretion of these growth factors begins within 10 min after clotting, and more than 95% of the presynthesized growth factors are secreted within 1 h (Fig. 10.1). The addition of thrombin and calcium chloride activates platelets in PRP and induces the release of factors from alpha granules.

Adult mesenchymal stem cells, osteoblasts, fibroblasts, endothelial cells, and epidermal cells typically express cell membrane receptors to growth factors present in PRP so these ones bind their transmembrane receptors, inducing an activation of internal signal protein. These processes cause cellular proliferation, matrix formation, osteoid production, collagen synthesis, etc.

It is important to underline that the PRP growth factors are not mutagenic because they don't enter in the cell or nucleus; therefore, PRP does not induce tumor formation.

10.2 Preparation

The phases of the working procedure of platelet gel therapy are collection into a test tube, cell enrichment, activation, quality control test, and record.

10.2.1 Collection into a Test Tube

PRP is obtained from a sample of patients' blood drawn at the time of treatment. A 40 cc venous blood draw will yield 7–9 cc of PRP depending on the baseline platelet count of an individual, the device used, and the technique employed.

G. Fabbrocini (✉) • M.C. Annunziata • C. Mazzella
Section of Dermatology, Department of Clinical Medicine and Surgery, University of Naples Federico II, Via Pansini 5, Naples 80131, Italy
e-mail: gafabbro@unina.it

S. Misso
UOC Medicina Trasfusionale e Immunoematologia – ASL Caserta, Caserta, Italy

G. Fabbrocini et al. (eds.), *Nonsurgical Lip and Eye Rejuvenation Techniques*,
DOI 10.1007/978-3-319-23270-6_10

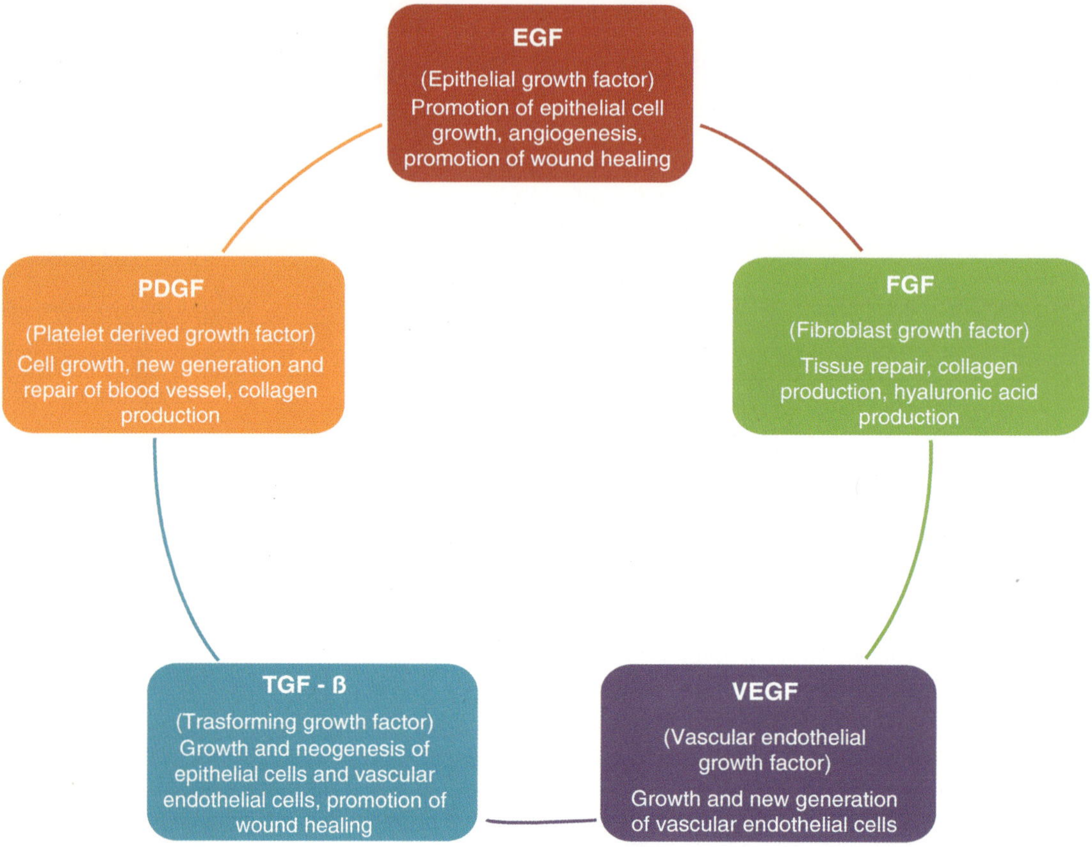

Fig. 10.1 Main growth factors present in PRP

The blood draw occurs with the addition of an anticoagulant, such as acid citrate dextrose A (ACD), to prevent platelet activation prior to its use. PRP is prepared by a process known as differential centrifugation. In differential centrifugation, acceleration force is adjusted to sediment certain cellular constituents based on different specific gravity.

There are many ways of preparing PRP. It can be prepared by the PRP method.

In the PRP method, an initial centrifugation to separate red blood cells (RBCs) is followed by a second centrifugation to concentrate platelets, which are suspended in the smallest final plasma volume. WB (whole blood) is initially collected in tubes that contain anticoagulants. The first spin step is performed at constant acceleration to separate RBCs from the remaining WB volume. After the first spin step, the WB separates into three layers: an upper layer that contains mostly platelets and WBC, an intermediate thin layer that is known as the buffy coat and that is rich in WBCs, and a bottom layer that consists mostly of RBCs. For the production of pure PRP (P-PRP), upper layer and superficial buffy coat are transferred to an empty sterile tube. For the production of leukocyte-rich PRP (L-PRP), the entire layer of buffy coat and few RBCs are transferred. The second spin step is then performed. "g" for second spin should be just adequate to aid in formation of soft pellets (erythrocyte-platelet) at the bottom of the tube. The upper portion of the volume that is composed mostly of PPP (platelet-poor plasma) is removed. Pellets are homogenized in 5 ml of plasma to create the PRP (platelet-rich plasma) high concentration of leukocytes.

10.2.1.1 PRP Method

1 Obtain WB by venipuncture in acid citrate dextrose (ACD) tubes.
2 Do not chill the blood at any time before or during platelet separation.
3 Centrifuge the blood using a "soft" spin.
4 Transfer the supernatant plasma containing platelets into another sterile tube (without anticoagulant).
5 Centrifuge tube at a higher speed (a hard spin) to obtain a platelet concentrate.
6 The lower 1/3 is PRP and upper 2/3 is platelet-poor plasma (PPP). At the bottom of the tube, platelet pellets are formed.
7 Remove PPP and suspend the platelet pellets in a minimum quantity of plasma (5–7 mL) by gently shaking the tube.

10.2.1.2 Buffy Coat Method

1 WB should be stored at 20–24 °C before centrifugation.
2 Centrifuge WB at a "high" speed.
3 Three layers are formed because of its density: the bottom layer consisting of RBCs, the middle layer consisting of platelets and WBCs, and the top PPP layer.
4 Remove supernatant plasma from the top of the container.
5 Transfer the buffy coat layer to another sterile tube.
6 Centrifuge at low speed to separate WBCs or use leukocyte filtration filter.

10.2.1.3 Commercially Available PRP Kits

There are many PRP systems commercially marketed, which facilitate the preparation of ready to apply platelet-rich suspensions in a reproducible manner. All operate on a small volume of drawn blood (20–60 mL) and on the principle of centrifugation. These systems differ widely in their ability to collect and concentrate platelets depending on the method and time of its centrifugation. As a result, suspensions of different concentrations of platelets and leukocytes are obtained. Differences in the concentrations in platelets and WBCs influence the diversity of growth factor concentration. It is difficult to assess which kit for PRP preparation is better and which is worse.

PRP devices can be usually divided into lower (2.5–3 times baseline concentration) and higher (5–9 times baseline concentration) systems. The high-yielding devices include Biomet GPS II and III (platelet count 3–8×), Harvest SmartPrep 2 APC+ (4–6×), and Arteriocyte-Medtronic Magellan (3–7×). The lower concentration systems include Arthrex ACP (2–3×), Cascade PRP therapy (1–1.5×), and PRGF by Biotech Institute, Vitoria, Spain (2–3×), and Regen PRP (Regen Laboratory, Mollens, Switzerland).

10.2.2 Cell Enrichment

In our experience, in order to prepare a gel that is a homogeneous mass of an adequate volume and yet remains manageable, the platelet concentration needs to be 750,000–1,000,000/μL. With this concentration of platelets, the gel forms in about 5–7 min. Once the PRP has been obtained, a full blood count is performed, and on the basis of the platelet count, the PRP is diluted or concentrated under sterile conditions.

10.2.3 Activation

The production of autologous thrombin, used as the activator, involves the following steps: collection of another blood sample (in ACD or sodium citrate), centrifugation of the sample for 10 min at 3000 rpm, collection of the plasma supernatant in a new test tube (under sterile conditions), addition of 0.2 mL of calcium gluconate for every 1 mL of plasma, incubation at 37 °C for 15–30 min, collection of the supernatant containing the precursors of thrombin (under sterile conditions), and freezing and storage at 30 °C until needed. In order to produce the gel, the platelet concentrate is placed in a sterile plate and then the activators are added, i.e., 1 mL of autologous thrombin and 1 mL of calcium gluconate for every 10 mL of PRP. At this point, the mixture is left to incubate at room temperature. If the coagulation process takes longer than expected, the preparation can be incubated for about 5 min at 37 °C to facilitate the reaction.

10.2.4 Controls of Quality and Sterility

The platelet concentrate must be sterile. The blood components must be prepared according to the principles of good manufacturing practices. Each procedure must undergo quality control tests including determination of the volume, platelet count, count of contaminating white blood cells, and assay of fibrinogen levels.

10.2.5 Records

The final product must carry a label indicating the surname and name of the patient, his or her date of birth, type of product, and the date of its preparation. The patient's personal data and the characteristics of the component are also recorded in the related files stored in the archives of the transfusion center. Complete physical examination and an analysis of the following clinical information should be obtained: general conditions of hygiene, lifestyle (smoking, alcohol), availability of family support, ability to walk, presence of occlusive arterial disease, past history of deep vein thrombosis, and presence of pain while walking and/or at rest (standing and/or lying).

10.3 Safety of PRP

Thanks to its autogenous preparation, PRP is safe and therefore free from transmissible diseases (HIV, HBV, HCV, West Nile fever, Creutzfeldt-Jakob disease); so, it is well accepted by patients.

Since the gel is homemade, it is probably a cheaper source of growth factors than the industrially produced ones and also provides growth factors not otherwise available for clinical use. The patients show good compliance toward the product and the procedures necessary for its production.

10.4 Properties of PRP

Angiogenic; antibacterial properties; stimulating the formation of connective and epithelial tissue; osteogenesis stimulators; security (219/05 Law, DD. MM. 3/3/05); nontoxic tissue injury; can be autologous; quick and easy preparation

10.5 Situations That Prevent the Production of Gel-Autologous Plt

Patient thrombocytopenic; vascular access compromise; septic patient; very large lesion; patient too small; patient positive for HBV, HCV, and HIV; emergencies. There are two types of contraindications to treatment with platelet gel: (1) those potentially harmful to the patient, such as hemodynamic instability, pregnancy, malignancies, infections, and/or osteomyelitis at the site of application, and (2) those making the autologous product difficult to obtain or of poor quality, such as thrombocytopenia, platelet disorders, and treatment with drugs affecting platelet function and/or coagulation (e.g., oral anticoagulants, heparin, nonsteroidal anti-inflammatory drugs).

10.6 Clinical Uses of PRP

After centrifugation, the platelet and fibrin component of the blood (the top layer) is extracted and reinjected into the area of concern (Fig. 10.2).

In dermatology and cosmetic medicine, PRP has been used to treat:

- Venous and arterial leg ulcers.
- Diabetic foot ulcers.
- Pressure ulcers (bedsores).
- Skin graft donor sites.
- First- and second-degree thermal burns.
- Superficial injuries, cuts, abrasions, and surgical wounds.
- Hair loss disorders – PRP has been shown to reinvigorate dormant hair follicles and stimulate new hair growth.

- Posttraumatic scars – PRP combined with centrifuged fat tissue and fractional laser resurfacing improve cosmetic appearance of scars.
- Facial rejuvenation – PRP injections can treat wrinkles, photodamage, and discoloration in conjunction together with other treatment modalities.

In the chrono-aging processes, dermal fibroblasts play a key role, thanks to their interactions with keratinocytes, adipocytes, and mast cells. Besides, they are also the source of extracellular matrix, proteins, glycoproteins, adhesive molecules, and various cytokines and increase the activation of the fibroblast-keratinocyte-endothelium axis, maintaining skin integrity.

PRP, increasing the length of the dermo-epidermal junction, the amount of collagen, and the number of fibroblasts, can be considered an effective therapy for skin rejuvenation: PRP in fact induces keratinocyte and fibroblast proliferation and typically collagen production amplification, increasing dermal elasticity.

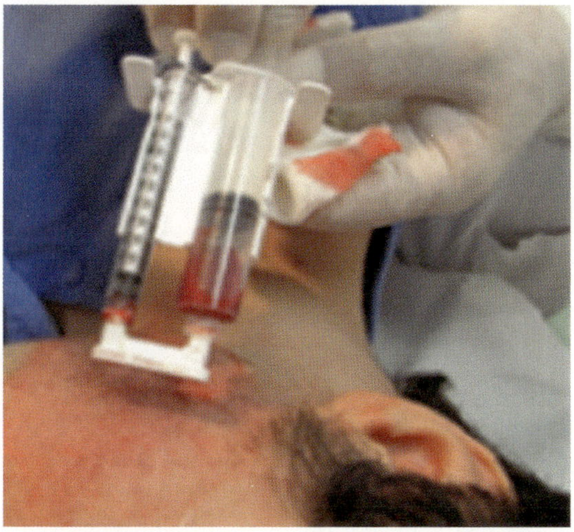

Fig. 10.2 PRP injection

It is also useful for tightening around the eyes (for thin crepe-like skin and fine lines) (Fig. 10.3) and in the areas as cheeks and midface, thinning skin on the neck, jawline and submalar regions, back of hands, décolleté, and others (e.g., knees, elbows, and upper arms, as well as for post-pregnancy skin laxity) (Fig. 10.4).

Besides platelet-rich plasma (PRP) can be used for enhancing, reshaping, and volumizing the lips; it is often used to improve very fine lines around the lips, helping to restore skin hydration and elasticity (Fig. 10.5).

A topical anesthetic or a nerve block will be used for pain management.

There are some additional effects using PRP combined with other aesthetic procedures as fractional laser or lipostructure. PRP in association with fractional laser increased skin elasticity and decreased the erythema index; keratinocyte and fibroblast proliferation and collagen production can explain these capacities.

The use of PRP mixed with purified fat graft has several advantages: PRP increased fat cell survival rate and stem cell differentiation.

This combination has been used for reconstructing the three-dimensional projection of the face contour in patients affected by facial aging characterized by atrophy of subcutaneous and soft tissue with loss of volume and elasticity, restoring the superficial density of facial tissue. PRP is an easily accessible source of growth factors for supporting bone and soft tissue healing.

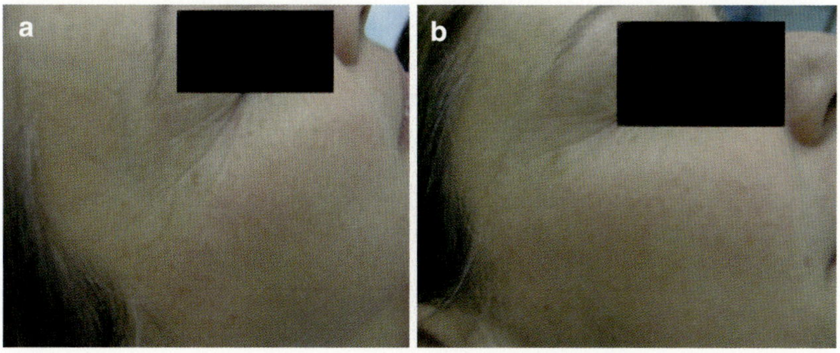

Fig. 10.3 Periocular area before (**a**) and after (**b**) PRP treatment

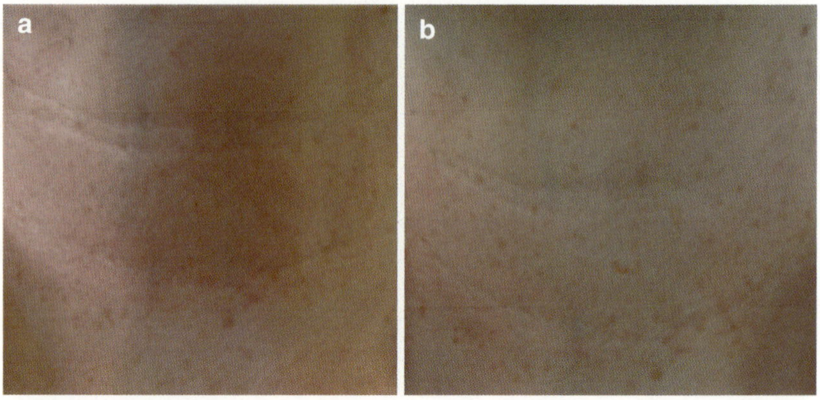

Fig. 10.4 Neck before (**a**) and after (**b**) PRP treatment

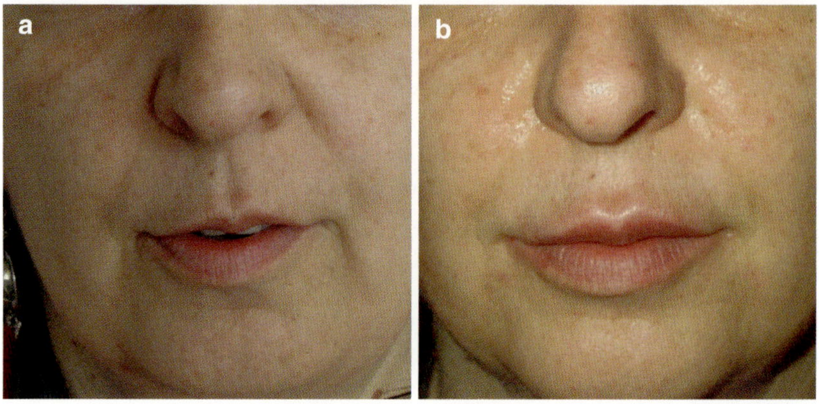

Fig. 10.5 Lips before (**a**) and after (**b**) PRP treatment

10.7 PRP and Platelet-Rich Fibrin Matrix (PRFM)

PRP can be enriched with the presence of a fibrin matrix (PRFM): fibrin matrices, in fact, enhanced the delivery of platelet growth factors. It consists of weak thrombin concentrations which entail equilateral junctions. These connected junctions permit the formation of a fine and flexible fibrin network capable of supporting cytokines and cellular migration that occurs. This results in an increase in the half-life of these cytokines as their release and use will occur at the time of initial scarring matrix remodeling. Thus, the cytokines are made available for a mandatory period required by the cells to initiate the healing.

Fibrin meshwork in PRF differs from that in PRP. In PRP, there are bilateral junctions resulting in a rigid network that does not honor the cytokine enmeshment and cellular migration. The increased thrombin required for rapid setting of the PRP leads to a rigid polymerized material.

PRFM has been proposed and effectively used in several facial plastic surgery settings: as PRFM can induce dermal augmentation, it can be used for treatment of dermal and subdermal tissues of the nasolabial folds, acne scars, and lip augmentation.

PRFM can be mixed with autologous fat ex vivo and the composite graft injected. The fibrin matrix associated with platelet-released growth factors should promote better graft take. This technique has been used for lip augmentation, with good results.

Bibliography

Cervelli V, Palla L, Pascali M, De Angelis B, Curcio BC, Gentile P (2009) Autologous platelet-rich plasma mixed with purified fat graft in aesthetic plastic surgery. Aesthetic Plast Surg 33(5):716–721. doi:10.1007/s00266-009-9386-0)

Cervelli V, Bocchini I, Di Pasquali C, De Angelis B, Cervelli G, Curcio CB, Orlandi A, Scioli MG, Tati E, Delogu P, Gentile P (2013) P.R.L. platelet rich lipotransfer: our experience and current state of art in the combined use of fat and PRP. Biomed Res Int 2013:434191. doi:10.1155/2013/434191

Dhurat R, Sukesh M (2014) Principles and methods of preparation of platelet-rich plasma: a review and author's perspective. J Cutan Aesthet Surg 7(4):189–197. doi:10.4103/0974-2077.150734

Fabi S, Sundaram H (2014) The potential of topical and injectable growth factors and cytokines for skin rejuvenation. Facial Plast Surg 30(2):157–171. doi:10.1055/s-0034-1372423

Kim DH, Je YJ, Kim CD, Lee YH, Seo YJ, Lee JH, Lee Y (2011) Can platelet-rich plasma be used for skin rejuvenation? Evaluation of effects of platelet-rich plasma on human dermal fibroblast. Ann Dermatol 23(4):424–431. doi:10.5021/ad.2011.23.4.424

Mehryan P, Zartab H, Rajabi A, Pazhoohi N, Firooz A (2014) Assessment of efficacy of platelet-rich plasma (PRP) on infraorbital dark circles and crow's feet wrinkles. J Cosmet Dermatol 13(1):72–78. doi:10.1111/jocd.12072

Misso S, Paesano L, D'onofrio M, Fratellanza G, D'Agostino E, Feola B, Minerva A (2006) Salvatore formisano: "our experience in the treatment of refractory ulcers with platelet gel". Blood Transfus 4:195–205

Sujeet Vinayak Khiste and Ritam Naik Tari (2013) Platelet-rich fibrin as a biofuel for tissue regeneration. ISRN Biomaterials. 2013:Article ID 627367, 6 page. http://dx.doi.org/10.5402/2013/627367)

Shin MK, Lee JH, Lee SJ, Kim NI (2012) Platelet-rich plasma combined with fractional laser therapy for skin rejuvenation. Dermatol Surg 38(4):623–630. doi:10.1111/j.1524-4725.2011.02280.x

Yuksel EP, Sahin G, Aydin F, Senturk N, Turanli AY (2014) Evaluation of effects of platelet-rich plasma on human facial skin. J Cosmet Laser Ther 16(5):206–208. doi:10.3109/14764172.2014.949274

11 The Nonsurgical Thread Lift for Facial Rejuvenation

Roberta Lovreglio, Gabriella Fabbrocini, and Mario Delfino

11.1 Introduction

Barbed suture lifting is a minimally invasive surgical technique for facial rejuvenation. Aging of the face and neck results in ptosis of soft tissues and the appearance of more prominent facial lines. For correction of these changes, surgeons are devising more procedures with fewer incisions and shorter postoperative recovery periods. Many of these procedures use absorbable and nonabsorbable sutures in the dermis and subcutis to lift lax skin. Limitations of these implants have included the protrusion of sutures through the skin and asymmetry of the cosmetic effect, often requiring correction with additional sutures, and limited durability of effects.

The treatment by absorbable threads, known as balance lift or nonsurgical face lifting, is an innovative technique used in aesthetic medicine that is useful for supporting and stretching the face and body tissues. The suspension threads are used to improve eyelid ptosis and the periorbicular groove, with a significant improvement of the shape of the eyebrows. They are used also to reduce ptosis of the neck, and middle and lower face (Figs. 11.1a–c and 11.2a–c). Aging induces a scaffolding of dermal facial skin as well as a progressively decreasing fat component, attributable to thinner connective tissue and collapse of elastic fibers. The affected areas generally include the cheeks, eyebrows, mandibular area, and neck. Dermatochalasis of the facial and neck soft tissues, including the superficial muscular aponeurotic system (SMAS) and the muscular tissue, is the cause of the distinctive aging signs on the face. The profile of the mandibular margin becomes unclear, displaying its deterioration (descent of the aging jaw line); the forehead has horizontal wrinkles where other vertical ones are added to the glabellar area; the zygomatic malar region (middle face) displays a downward trend; a lachrymal furrow appears and the nasobuccal and buccomandibular areas deepen; the skin of the eyelid becomes flabby and protrudes in correspondence with the lower eyelid, owing to production of adipose bubbles; and finally, a plasmatic parcel and cutaneous flabbiness appear on the neck. The facelift to correct facial aging has evolved into an elaborate and complicated procedure requiring a lengthy recovery time. The recent introduction of absorbable barbed sutures producing a lifting action for this type of aging offers a good alternative to more invasive procedures. The plugs present on

R. Lovreglio, MD (✉) • G. Fabbrocini, MD
M. Delfino, MD
Division of Clinical Dermatology,
Department of Clinical Medicine and Surgery,
University of Naples Federico II,
Via Sergio Pansini 5, 80133 Napoli, Italy
e-mail: robertalovreglio@gmail.com;
gafabbro@unina.it

G. Fabbrocini et al. (eds.), *Nonsurgical Lip and Eye Rejuvenation Techniques*,
DOI 10.1007/978-3-319-23270-6_11

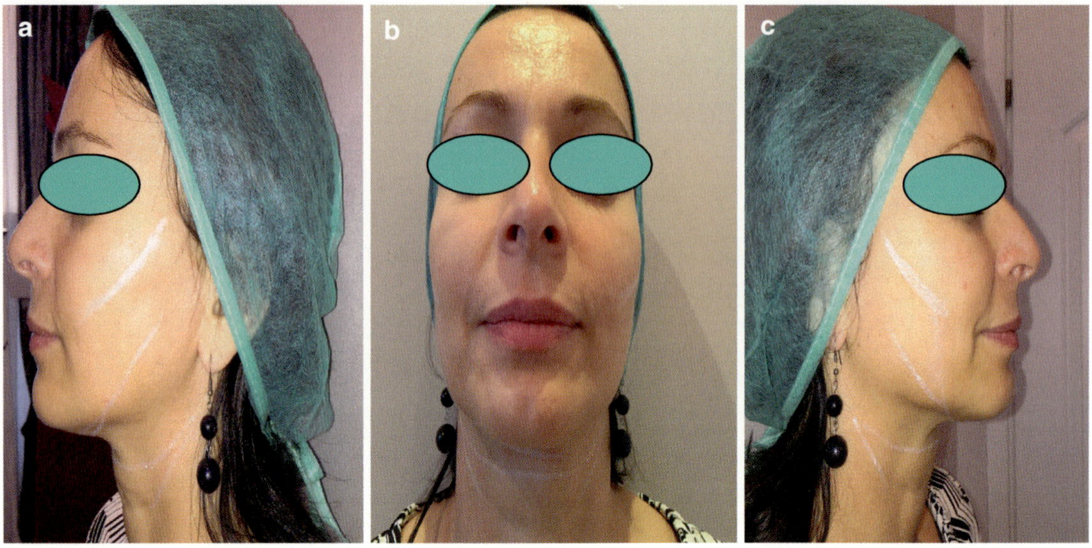

Fig. 11.1 (**a–c**) Before treatment

the surface of the wires allow the combination with other nonsurgical rejuvenation procedures, such as botulinum toxin or substances with a transient and volumizing filler effect. The implant can be performed in an outpatient setting under local anesthesia. The surgeon first establishes the degree and direction of the desired tightening. This determines the course and number of sutures that have to be placed to achieve the best result. Infiltration of local anesthesia is limited to these lines and the insertion points of the straight needle. For lifting of the brow and middle and lower face, 3- to 4-mm incisions for insertion of the straight needle are made posterior to the frontal and temporal hairline. For lifting of the neck, incisions are made posterior to the sternocleidomastoid muscle of the lateral neck. To place an individual thread, the surgeon guides the straight needle through the incision and into the subcutaneous plane. For some anatomic locations it is advantageous to bend the needle to more easily allow it to follow the dynamic face lines. The needle is advanced in this plane in a zig-zag movement along the marked trajectory. Once anchored, this zig-zag placement of the suture limits retrograde motion along the suture and results in an implanted suture that is longer than the drawn trajectory. This maximizes the number of barbs in the subcutis and theoretically provides greater stability of the translocated skin. Movement of the needle and suture through the subcutis is generally well tolerated by patients. If the straight needle moves superficially to this plane, it is immediately apparent as linear dimpling of the overlying skin. If the needle enters into the deep subcutis or approaches the muscle fascia or periosteum, the patient will report the sensation of pain or pressure. At any point, the straight needle may be partially or completely removed and repositioned. The straight needle exits the skin inferior to the eyebrow or near the medial face or neck. It is then cut from the thread after pulling the attached suture through the skin. This leaves the barbed portion of the thread buried in the subcutis with the free ends extending from the proximal insertion point and the distal medial face exit point. The curved needle on the proximal end of the suture may then be used to anchor the suture near its insertion to the underlying fascia or periosteum. A 3- to 4-mm incision, 1–2 cm posterior to the insertion points, serves as an exit point for the curved needle after deep suturing to the fascia or periosteum. Greater security of this anchor point is achieved by tying this suture at its proximal end with a paired suture running a similar parallel course in the skin. The resulting knot can be seated in this posterior incision by

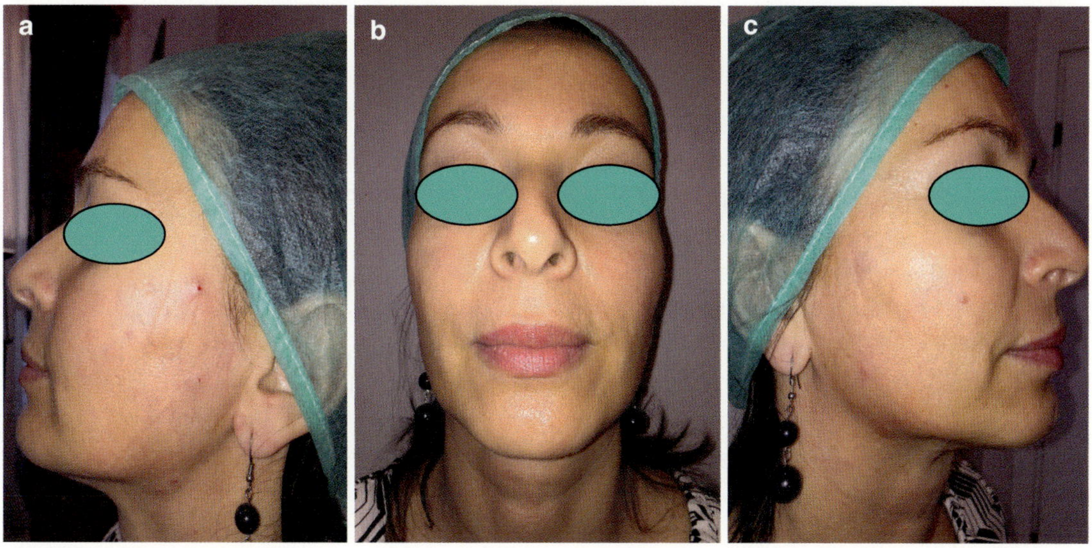

Fig. 11.2 (**a**–**c**) After treatment

gentle traction on the distal ends of the paired sutures. When all planned sutures have been placed and anchored, the patient returns to the seated position. Holding the distal end of the suture protruding from the medial face, brow, or neck with one hand, the surgeon uses the other hand to push the lax skin overlying the suture toward the anchoring point. The unidirectional barbs catch on the fibrous septae of the subcutis, preventing retrograde movement. Together, the surgeon and patient decide the degree of tightening along any given suture. The distal end of the suture extending from the medial face, brow, or neck is cut at its exit point and retracts under the skin. Incisions used for insertion of the straight needle and anchoring heal rapidly by secondary intention. The translocation of the skin along the suture can cause lax skin folds in the hairline and lateral neck that can be quickly and completely remodeled or redistributed to the scalp and neck in several days or weeks. Mild complications such as swelling, bruising, and subjective feelings of "tightness" usually resolve within 1–3 weeks. Transient neuropathy of the greater auricular nerve has occurred in several patients when using the sternocleidomastoid muscle fascia as an anchoring point on the lateral neck. Because this technique may be released with intense pressure, patients must initially avoid strong exercise or movements that could dislodge the tightened skin from the hundreds of barbs along the sutures. Non-peer-reviewed data from the manufacturer demonstrate that in laboratory rats these sutures develop a fibrous capsule that becomes well integrated into the dermis and subcutaneous tissue over several months. A similar process in human skin can lead to a long-lasting cosmetic effect. The actual long-term durability of the tightening effects of these sutures is unknown. Early adopters of this procedure have demonstrated maintenance of cosmetic effects at 6 months. The technique will lead to a true firming of the skin affected by laxity if the thread is placed along the traction lines. The threads are located in a direction perpendicular or at an obtuse angle with respect to the traction lines of the skin, to obtain a tightened effect. The number of wires implanted may vary according to the material of the thread used and the form chosen by the operator. A worldwide variety of barbs, with differences in costs and types of materials, are available. Here we list our experience of the main ones:

1. PDO – Polydioxanone
2. Polylactic acid – Poly(L-lactide)-ε-caprolactone copolymer

11.1.1 When Using the Threads

The application of suspension threads can be used at the face level in the following cases:

- Loose and heavier jaw line
- Soft tissues descending to cheekbones and cheeks level, with a consequent loss of natural volume and fatigue of the face
- Presence of pronounced furrows and wrinkles;
- Breakdown of the skin in the neck area, mainly in the central zone
- Lifting the eyebrow arch

Suspension threads can be used at the body level in the following cases:

- Tissue laxity in the inner arms
- Firming the abdomen
- Skin breakdown inside the thigh
- Sagging skin in the upper area of the knees;
- Loss of form at the navel due to partial coverage by the abdominal skin
- Mild ptosis of the breasts
- Soft ptosis of the buttocks

11.1.2 Absolute Contraindications

- Current acute acne or skin diseases
- Systemic infections
- Allergic anamnesis
- Treatment with immune suppressors
- Ongoing cancer and liver treatment
- Uncontrolled hypertension or anticoagulant regimes

11.2 PDO Threads

Nature PDO (polydioxanone) is a polymer of polylactic caprolactone, a bioabsorbable material and antimicrobial already used in surgery and biocompatible with the dermis.

Recommended Applications The indication is for sagging and relaxation, or when preventive treatment is needed.

Features Biocompatible, antimicrobial, Absorption by hydrolytic action within 6–8 months, suitable for medical use and surgery

This material is already used in general and specialized surgery as sutures; it is slowly absorbable, biocompatible, and antimicrobial.

Modalities of Implant The threads are implanted in the skin. In the preorbicular area, they are inserted in a direction along the front lines and/or along the brow for lifting, to improve periocular dermatochalasis. They are inserted obliquely by hand at the side of the eye down to within 1 cm of the medial canthus. At the level of the labia, they are inserted to follow along the labia for lip contouring, along the vermilion for improving the firmness of the lips, and at the corners of the lips in a direction perpendicular to the wrinkles to enhance the face expression.

Functions Generally speaking, the shift of the skin tissue through the thread represents a fundamental mechanism in cutaneous tissue repair processes and tissue regeneration. The cell is stimulated by the presence of several transduction systems mechanically placed at the level of the cell membrane, the best known of which is integrin. The mechanical stimulus exerted on the outer extracellular matrix determines via integrin the intracellular biological changes that can activate specific genes. The fibroblasts are particularly sensitive to mechanical stimuli, and when subjected to mechanical stimulation they activate genes for the production of collagen and other proteins. The biostimulant PDO threads implanted in the dermis are able to stimulate the fibroblasts, activating an increased synthesis of collagen. The PDO absorbable threads (thickness 0.05–0:19 mm and length 3–16 cm) are positioned inside a needle (26, 29, or 30 gauge). After a preliminary visit the clinician must accurately mark with a dermographic pen the face areas, neck, and décolleté that must be treated, according to the situation and the demands of the patient. Insertion of the needles (29–31 gauge) with PDO threads can then take place without trauma and without any need for local anesthesia. The needle tips used in this technique are shaped to reduce pain. The surface of the needle also has a double coating that makes it homogeneous. The number of threads released into tissues during treatment varies depending on several factors such as patient's age, degree of skin aging, degree of failure of the tissues, and extent of the area to be treated. Usually 20–60 threads are used. The needle is inserted fully and advanced, leaving a thread inserted and implanted in the tissues. Anesthesia is not necessary; cooling the treated area with dry ice will be sufficient. In the case of more than 10 implants, we recommended a bandage (e.g., Tensoplast) to restrict movement in the first 24 h postoperatively.

11.2.1 After Treatment

Rapid lifting of the treated area is visible immediately after the end of treatment. The patient can restart daily activities at the end of treatment. The thread is inserted in directions orthogonal to each other to form a grid that counteracts the gravitational vectors that lead to relaxation, behaving like a foreign body coated by collagen, which forms the true support structure desired. PDO has been used for decades in surgery for the execution of absorbable sutures.

11.2.2 After 2 Months

There is a marked improvement after 1–2 months. The skin tone also improves, as does the appearance of wrinkles.

Results and Benefits An innovative technique in the field of aesthetic medicine, with very little pain, which improves the skin tone and its aspect.

Duration after approximately 6–8 months, the threads in the PDO have been completely reabsorbed by hydrolytic action in a totally natural and harmless manner, but the biostimulation and lifting effect continues the mechanical support and remains stable for 6–8 months, and generates a significant stimulation of endogenous cells whose benefits will last much longer.

Adverse Reactions Edema and erythema for 24–48 h after treatment, hematomas at implant, hardening small transients.

High sensitivity in the treated area, which usually disappears within 1–3 days following treatment

Removal of the Threads In extreme cases the threads can be extracted, within 20 days after treatment, by a small incision and extraction using forceps.

Contraindications Skin infections, recurrent herpes simplex (requires antiviral prophylaxis), collagen, predisposition to form keloids, pregnancy, lactation, coagulation disorders, exaggerated expectations of cosmetic surgery.

11.2.3 Types of Threads in PDO

- Monofilament
- Screw
- Derma Spring
- Barbed threads
- Bidirectional (Cog)
- Multidirectional (Cog)
- 4D Cog

11.3 Screw Threads

The Screw is the latest innovation in the field of non-invasive mini-lifting of the face and body. Thanks to its special shape the microthread support "screw" produces greater vascularity with improved macrophage response that makes more effective the biostimulative response, especially for fibroblasts, which become more "active" and more "flexible" in the healing process.

11.3.1 Applications of Reinforced Screw Face Threads

The combination between the microthread, a PDO monofilament, and the screw grants significant synergy, which results in better toning action on the connective tissue, as the thread screw stimulates mostly the "cascade effect" of the systems of healing and restructuring tissue.

11.3.2 Applications of Reinforced Screw Body Threads

Effective in activating the process of restructuring the dermoepidermal junction, the fine thread contour technique reshapes the body.

11.3.3 Security Reinforced Screw Threads

- Medical device class III with CE mark no. 1293
- Registration at the Medical Report Italian Ministry of Health no. 875113
- Certificate PCPC (Personal Care Product Council)
- Noninvasive, fast, safe
- Does not produce scars
- Minimal risk of hematoma
- Applicable on face and body
- Wires PDO 100 % biocompatible

11.4 Bidirectional and Multidirectional Barbed Threads

Barbed bidirectional threads are useful to create a "mini-invasive suspension surgery" in a treatment that allows lifting and repositioning of tissues with greater modeling. The barbed bidirectional threads bring tissue to the opposite direction through bidirectional pinning, with minimum risk of displacement or migration, thus facilitating the anchorage in the tissue and enhancing the lifting effect. Multidirectional threads are useful for all ptosis tissues located at the lateral third of the eyebrow, and all applicants requiring a traditional facelift and surgical and aesthetic facial improvement, as well as fillers and lipofilling

11.4.1 Features

- Soft lifting with vertical traction
- Faster processing
- More invasive than Screw
- Requires learning curve for proper technique
- Requires knowledge of surgical plan and sometimes small amount of local anesthesia

11.4.2 Safety

- Medical device Class III with CE mark 1293
- Registration at the Medical Report Italian Ministry of Health no. 875113
- Certificate PCPC (Personal Care Product Council)
- Minimally invasive, fast, safe
- Low risk of scars
- Minimal risk of hematoma
- Applicable on face and body
- Wires PDO 100 % biocompatible

Nowadays micropipes can be used as an alternative to microneedles and offer greater advantages:

- Less invasive
- Reduction of bleeding
- Less risk of vessel injury

The micropipes must be used carefully, and sometimes have been used to make microtunnels with a fine pipe under local anesthesia or products containing connective micrografts. Micropipes are recommended only for selected areas and methods. The micropipe Cogs multidirectional (21 gauge × 60–90 mm) are typically indicated for the frontal area as well as the neck. The lower eyelid (PDO 30 gauge × 27 mm) is very useful with this micropipe, allowing the placement of microthreads in the lower area of the eyelid. The Short pipe for nose and neck (19 gauge × 40 mm 4D thread) is used for correction of the silhouette of the nose in selected cases, by pulling on the tip and reducing mild disharmonies, creating an interesting volumetric effect. The 4D screw micropipes allow thread cutting in four dimensions. In this way they are able to optimize the lifting effect and minimize the damage to tissues.

11.4.3 Major Contraindications for All Types of PDO Threads

- Acute acne and ongoing skin diseases
- Systemic infections
- Allergies to materials
- Treatment with immune suppressors
- Liver cancers
- Uncontrolled hypertension in treatment of coagulopathies
- Patients with very pronounced excess skin

11.5 Polyactic Acid/Poly-ε-Caprolactone Threads

These threads are biocompatible and fully resorbable. Polylactic acid and caprolactone also have a revitalizing action and are reported to be long lasting.

11.5.1 History

Professor Marlen Sulamanidze, a specialist in reconstructive plastic surgery and aesthetics, is recognized by the worldwide medical community as the inventor in 1995 of the first wire for lifting tissue ptosis. In 2008 he launched on the world market the three types of resorbable wire Nano, Excellence, and Light Lift. He invented a noninvasive lifting technique using permanent traction threads in polypropylene. The technique has evolved over 8 years of experience with nonresorbable threads.

Patent Pending and Safety 16 patents over the years

11.5.2 Patient Types and Areas of Application

- Face
- Eyebrow
- Zygomatic area
- Chin area
- Subchin area
- Forehead wrinkles
- Glabellar lines
- Nasolabial area
- Folds wrinkles
- Chin and neck wrinkles
- Body
- Remodeling of the thighs (toning and lifting)
- Increased tonicity of décolleté
- Reduction of ptosis
- Firming and lifting of belly

This is a surgical outpatient treatment carried out with or without local anesthesia according to the medical evaluation of the case. Threads are inserted subcutaneously through a microhole, along precise lines of skin tension, by means of a thin needle or a needle-pipe blunt tip (tip shape reduces the local trauma), exerting a slight traction to lift the relaxed tissues. The threads adhere to the skin owing to the presence of special anchors (plugs). The effect is achieved because the lift-thread insertion follows geometric traction where nothing is left to chance, and in fact the treatment requires, in addition to a certain manual skill, a flawless knowledge of anatomy.

11.5.3 Material

The caprolactone allows the gradual absorption and uniformity of polylactic acid while ensuring the mechanical strength and the elasticity of the thread in time; also, the capacity of biostimulation is associated in time with the effect of traction of these threads in restoring luminosity and color.

11.5.4 Histology

The microcirculation capillaries of the proximal threads increase in number compared with the peripheral circulation, and the vessel lumen is more dilated. Studies have shown that for the entire period after insertion the vessels remain dilated, with constant hyperemia maintaining a trophism of the treated area with formation of new collagen, fibrin, and elastin. The fibroblasts of the fibrotic tissue created with the placing of the threads are functionally active, observable from increased spread of chromatin in the nucleus and cytoplasm. The connective tissue layer in which the thread is implanted, treated with the blue dye toluidine, contains an increased number of mast cells, and at the same time show an increased concentration in the vascular channels of the microcirculation. The granules of mast cells contain hyaluronic acid, a polysaccharide complex, which is a structural component of the papillary dermis granular layer of the epidermis and is also apparent in the superficial vessels of the skin. There is evidence that reduction of the amount of hyaluronic acid affects the immune status of the skin, and that its inner dermal injection improves the structure of the skin.

11.6 Histological Subskin Representation

11.6.1 Short Thread

- Thread typology of biowoven fibers with bulking revitalizing tensor effect
- Type of biofibers is spiral, with bulking, revitalizing, and firming effect for areas of greatest dynamism

11.6.2 Interval Reabsorption

The reabsorption process begins after about 180 days after implantation and is completed after more than a year. The unrolling of sutures takes placed within 3 weeks from the implant, since the thread is composed of two strands. The process of opening the thread by imbibition in the dermal tissue leads to an affection tensor characteristic unique to this type of thread, i.e., ready-to-use, "pre-filled" braided and sutured, 4 cm long, 23 gauge.

11.6.3 Zones of Application Can Be Multiple Ones

- Front
- Nose
- Cheekbone
- Décolleté
- Chin
- Neck
- Hands

Anchoring in the medium/deep dermis, on both the face and body, are soon remarkable. The results are apparent after 3 months with a progressive improvement in the following months.

11.6.4 Long Hypodermis Threads

11.6.4.1 Face

This thread is composed of a needle-pipe preloaded suture, characterized by a multidirectional microanchor of 12 cm with an indication for treatment of ptotic skin areas more pronounced in various areas of the face. It is best indicated for the lifting of the cheekbone.

11.7 Areas of Face Correction

- Eyebrows
- Glabellar wrinkles
- Submalar areas
- Marionette lines
- Nasolabial folds
- Mandibular area
- Chin
- Submaxillary area

11.7.1 Thread Indicated for the Correction of the Chin and Neck Area

Rubber thread with a double needle without a lifting skin retraction with a microanchored convergence.

11.8 Problematic Situations and Complications

- Thread breaking as a result of a loosening
- Emergence of the thread from the skin
- Dimples at the entry points of the thread
- Bruising
- Migration of thread
- Asymmetries
- Overcorrection
- Cutaneous retractions
- Inflammation
- Migration
- Other

11.9 Home Therapy

- If necessary, antibiotic therapy
- Therapy with anti-edema substances such as bromelain or diosmin

11.10 Treatment of Correction

- Physiotherapy 5 %
- Fillers 5 %
- Removing the suture 3 %
- Inflammation correction 2 %
- Removal for migration 2 %
- Needles for the removal of sutures

11.11 Recovery Time

Usually about 3–5 days. After surgery the patient should:

- Use cold compresses for 24 h
- Use antiseptic solutions for 3 days
- Not take food, hot liquids, and solids for 3 days
- Avoid alcohol for 2–3 weeks
- Limit mimicry activity for 7 days
- Limit gym, sauna, swimming, and exposure to direct sunlight for 3–5 weeks
- Use antibiotics for 3–5 days, in the case of lowered immunity or if using more packs of sutures for surgery
- Sleep supine or side to side with a pillow in the case of face, neck, and abdomen surgery

11.12 Contraindications

- Autoimmune diseases
- Collagenopathy
- Coronary heart disease
- Hypertension II and III
- Inflammation or cancer in the target area
- Propensity of keloids and hypertrophy
- Taking anticoagulants
- Pregnancy, breastfeeding
- Previous injectable biodegradable products in the area of the procedure
- Individual intolerance to medicines needed

After implantation the patient must avoid massaging the treated area, exposure to direct sunlight for a month or so, and saunas and gym for at least 3–5 weeks.

Conclusions

None of the complications presented herein has generated the appearance of long-term functional disorders and no visible and permanent effects. Moreover, none of the complications presented required the need for prolonged treatment. Innovations in operative techniques generally contribute to enhanced results, greater patient happiness, and a decrease in operative morbidity. The immediate effect is the lifting of the tissue, owing to the mechanical action produced by the thread, which contrasts with the falling of the area treated (Figs. 11.3a–c and 11.4). This is possible because of the arrangement of the threads'

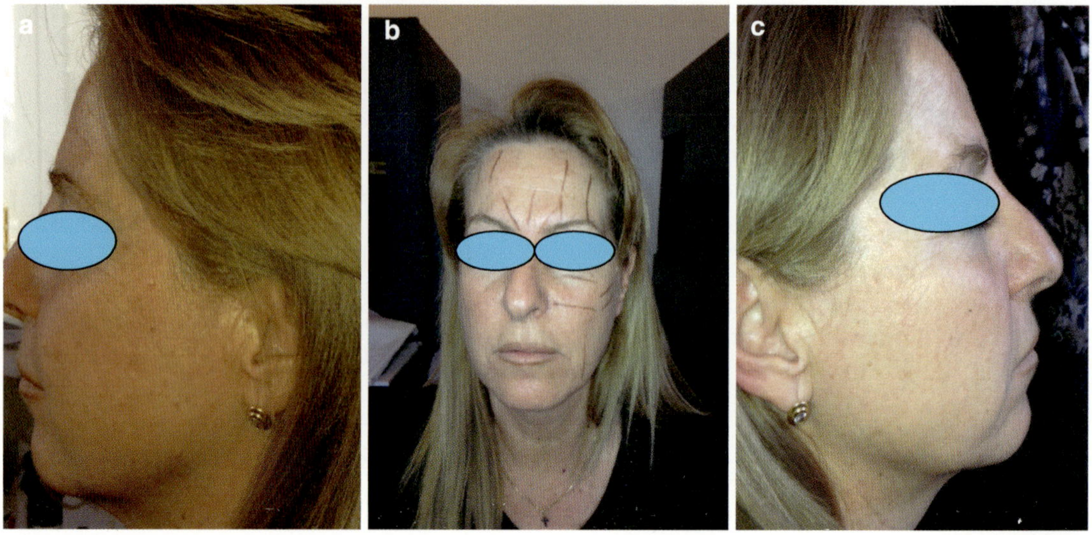

Fig. 11.3 (**a–c**) Before treatment

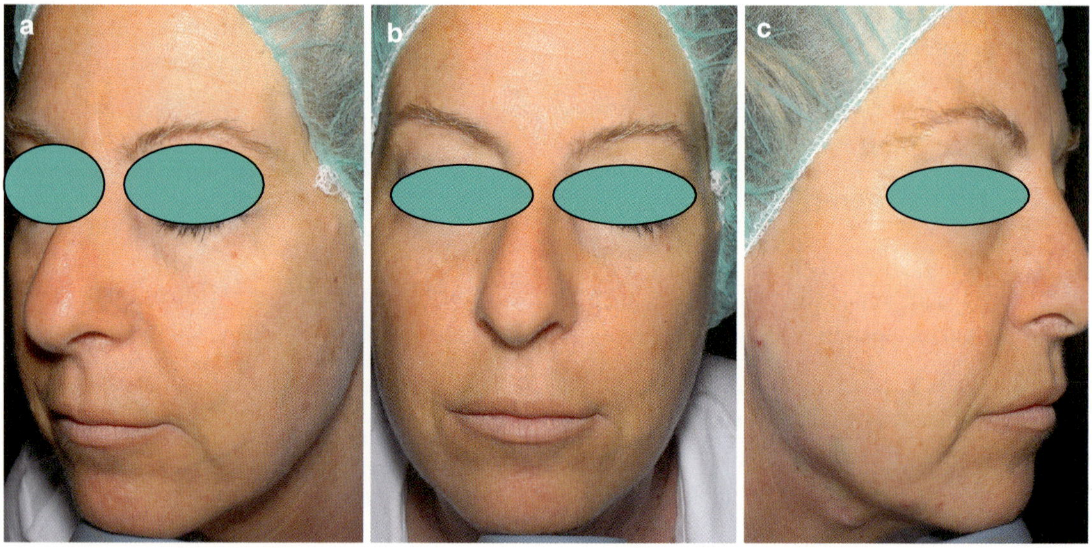

Fig. 11.4 (**a–c**) After treatment

barbs, disposed in two directions (divergent and opposite), compared with the middle point of the thread. Once positioned in the subcutaneous tissue, the threads will continue to exert their sustaining action on the tissues. Therefore, it is possible to claim that the lifting effect is guaranteed and fortified by the cutaneous reaction (fibrosis) that appears along the length of the thread, which remains effective and steady even when the thread is completely reabsorbed (after about a year). The reabsorption occurs as a result of the action of the histiocytic-reticule system, which concretizes a selective hydrolysis action of the reabsorbable thread from the periphery toward the center. The most important limits of this technique are that it is indicated for moderate cutaneous descent. For overabundant tissue, the prescription remains traditional lifting. In cases of more advanced and evident signs of aging, patients must opt for traditional surgical options that are more invasive and direct. Therefore, strict selection criteria must be adopted when selecting patients to be treated with this technique. Reabsorbable Happy Lift Revitalizing thread constitutes an efficient and safe procedure of mid-face lifting and rejuvenation of the superior cervical region of the face and neck. It is also possible to combine this with other methods that allow optimization of the facial rejuvenating effect, such as botulinum toxin, fillers, chemical peelers, photorejuvenation with a pulsed light, and lip filling. These threading procedures do not require general anesthesia, are virtually free of bleeding or pain, and do not produce intra- and postoperational scars that are visible on the skin, nor do they require long postoperative recovery times. The technique is practicable in day surgery, and the patient may immediately return everyday activities shortly following the procedure.

Bibliography

1. Chaffoo RA (2013) Complications in facelift surgery: avoidance and management. Facial Plast Surg Clin North Am 21:551–558
2. Giampapa VC, Di Bernardo BE (1995) Neck recontouring with suture suspension and liposuction: an alternative for the early rhytidectomy candidate. Aesthetic Plast Surg 19(3):21–23
3. Helfrich YR, Sachs DL, Voorhees JJ (2008) Overview of skin aging and photoaging. Dermatol Nurs 20:177–183
4. Lycka B, Bazan C, Poletti E, Treen B (2004) The emerging technique of antiptosis subdermal suspension thread. Dermatol Surg 30(1):41–44
5. Nkengne A, Bertin C (2012) Aging and facial changes– documenting clinical signs, part 1: clinical changes of the aging face. Skinmed 10:284–289
6. Sasaki GH, Cohen AT (2002) Meloplication of the malar fat pads by percutaneous cable-suture technique for mid-face rejuvenation: outcome study (392 cases, 6 years' experience). Plast Reconstr Surg 110(2):635–654
7. Silva-Siwady JG, Diaz-Garza C, Ocampo-Candiani J (2005) A case of Aptos thread migration and partial expulsion. Dermatol Surg 31(3):356–358
8. Sulamanidze MA, Sulamanidze G (2008) Facial lifting with Aptos methods. J Cutan Aesthet Surg 1(1):7–11. doi:10.4103/0974-2077.41149
9. Sulamanidze MA, Sulamanidze G (2009) APTOS suture lifting methods: 10 years of experience. Clin Plast Surg 36(2):281–306. doi:10.1016/j.cps.2008.12.003, viii
10. Sulamanidze MA, Fournier PF, Paikidze TG, Sulamanidze GM (2002) Removal of facial soft tissue ptosis with special threads. Dermatol Surg 28(5): 367–371
11. Sulamanidze MA, Paikidze TG, Sulamanidze GM, Neigel JM (2005) Facial lifting with "APTOS" threads: featherlift. Otolaryngol Clin North Am 38(5): 1109–1117
12. Sulamanidze MA, Sulamanidze G, Vozdvizhensky I, Sulamanidze C (2011) Avoiding complications with Aptos sutures. Aesthet Surg J 31(8):863–873
13. Villa MT, White LE, Alam M, Yoo SS, Walton RL (2008) Barbed sutures: a review of the literature. Plast Reconstr Surg 121:102e–108e

Complications of Hyaluronic Acid Fillers

12

Raymond Fertig, Maria Pia De Padova, and Antonella Tosti

Hyaluronic acid dermal fillers have become a mainstay for soft-tissue augmentation while providing numerous advances in the area of cosmetic surgery. HA fillers are primarily used for the treatment of facial changes associated with aging, which include thinning of the epidermis, loss of skin elasticity, subcutaneous fat and bony changes, and atrophy of muscle, all of which can result in a loss of volume.

HA fillers are longer lasting and less immunogenic, making them the most common of the temporary fillers on the market. The vast majority of treatments are efficacious and patient satisfaction is generally high. Despite having a low overall side effect profile, early and delayed complications, ranging from minor to severe, have been reported following HA filler injection. The most common potential sequelae following HA filler injection result from the injection site reactions and include ecchymosis, edema, erythema, and pain. Other less common adverse events include nodules (both inflammatory and noninflammatory), hyperpigmentation, telangiectasia, and dyschromia. More serious adverse events, while rare, include vascular compromise that can result in tissue necrosis and acute vision loss.

R. Fertig
Department of Dermatology and Cutaneous Surgery, University of Miami, Coral Gables, FL, USA

M.P. De Padova (✉)
Department of Dermatology, Ospedale Privato Accreditato Nigrisoli, Bologna, Italy
e-mail: mdepadova@gmail.com

A. Tosti, MD
Department of Dermatology and Cutaneous Surgery, University of Miami, Coral Gables, FL, USA
e-mail: ATosti@med.miami.edu

12.1 Injection-Associated Pain

Some degree of pain is expected with needle puncture, with thicker gauge needles expected to cause a greater degree of pain due to more extensive tissue injury. Where the injection is placed can also determine how much pain is experienced, as more sensitive areas tend to be more painful such as injections of the lip, injections of the periocular skin, and perioral injections. Injection site pain is minimized by the formulation of hyaluronic acid fillers to include lidocaine. If the HA filler does not include an anesthetic, a topical anesthetic ointment can be applied before treatment. The topical anesthetic should be applied 20–30 min before injection. In addition, cold compresses can be applied just before injection to numb the area to diminish pain sensation.

12.2 Skin Discoloration

12.2.1 Erythema

Erythema (i.e., redness) is frequently observed immediately after injection with HA fillers.

G. Fabbrocini et al. (eds.), *Nonsurgical Lip and Eye Rejuvenation Techniques*,
DOI 10.1007/978-3-319-23270-6_12

Erythema is a local effect due to puncture trauma and associated inflammation. Erythema is best managed by applying cold compresses postinjection for 5–10 min to reduce inflammation. After the procedure, patients can be advised to use ice packs at home every few hours on the day of the injection. Caution must be emphasized to avoid prolonged ice pack usage to reduce the risk of cold injury to the skin. In addition, vitamin K cream can be useful in accelerating resolution of erythema. Furthermore, redness can be reduced using a prudent injection technique that will minimize the number of skin punctures during the injection process, thus limiting trauma to the injected area. Such techniques include placing the filler with a serial injection, using the fanning technique, or using linear tunneling with threading.

The erythema will generally last for a few hours and may persist for a few days. If the erythema persists longer than the expected duration of a few days, then a hypersensitivity reaction is suspected. Effective treatments for this hypersensitivity reaction include oral tetracycline or isotretinoin [1]. For persistent erythema, a medium-strength topical steroid is warranted. High-potency steroids should not be used as they increase the risk of atrophy and telangiectasia [1]. Of note, patients with rosacea have a higher risk of developing postinjection erythema and should be warned of this risk prior to commencing treatment [1].

12.2.2 Ecchymosis

While not as common as redness and swelling, ecchymosis (bruising) is an adverse effect that can occur in patients after receiving a hyaluronic acid filler injection. Bruising is caused by the perforation of vessels during filler injection, typically the dermal veins. Additionally, pressure of the injected material can cause injury to proximal blood vessels, causing bruising. Common locations for bruising are the upper third of the nasolabial fold, the upper lip, the lateral edge of the lower lip, and the perioral region. In particular, injections to the lower eyelid often result in bruise formation. Bruising is frequently observed after injection into the dermal and immediate subdermal planes using fanning and threading techniques [2].

Bruising may develop soon after the injection, but often is delayed, most notably in those patients who are on anticoagulation therapy. Therefore, patients should be counseled to discontinue any unnecessary anticoagulation medications or products 1 week prior to treatment to potentially reduce the severity of bruising. The blood-thinning products to be avoided include aspirin, nonsteroidal anti-inflammatory drugs (NSAIDs), warfarin, clopidogrel, dipyridamole, garlic tablets, ginkgo biloba, ginseng, fish oil, St. John's wort, and vitamin E supplements [3]. Bruising can be mitigated by applying cold compresses and firm pressure to the affected area before and after the procedure. Furthermore, vitamin K cream can be useful in accelerating the healing of bruising, just as it is for erythema [4]. Laser treatment may also accelerate the elimination of a bruise. Furthermore, postinjection bruising may be limited in extent by the incorporation of epinephrine in the filler which causes vasoconstriction and dampens the activity of eosinophils that can cause bruising [1]. Other recommendations to limit bruising include using the smallest gauge needle possible that can effectively deliver the filler, delivering small aliquots of product utilizing a slow injection technique, using the depot technique at the preperiosteal level and limiting the number of transcutaneous puncture sites [1]. The use of blunt cannulas may also limit bruising [5].

The typical time course for resolution of a bruise is approximately 1 week and can range from 5 to 10 days. Concerned patients should be instructed that the bruise may progress to a darker discoloration for 1–3 days posttreatment before it slowly resolves over 5–10 days. Patients should be made aware that the development of a bruise will not interfere with the treatment outcome.

12.2.3 Telangiectasia

Telangiectasia is an abnormal aggregation of arterioles, capillaries, or venules. This neovascularization process is an adverse outcome that may occur at the HA filler injection sight. The proliferation of these vessels is caused by trauma to the tissue due to the product causing tissue expansion [6]. Telangiectasia can appear within days or weeks following the procedure. Left untreated, they typically resolve within 3–12 months [6]. Telangiectasias following dermal filler injection have been successfully treated using a 532-nm laser (532 nm KTP and the 532 nm diode copper vapor) or 1,064 nm laser [7]. Other forms of effective laser therapy for telangiectasias include intense pulsed light (IPL) and 585 nm pulsed dye laser [6]. In addition to laser treatment, telangiectasias can also be treated with hyaluronidase [8].

12.2.4 Hyperpigmentation

The trauma induced by dermal filler procedures, including HA dermal injection, can cause post-inflammatory hyperpigmentation. Post-inflammatory hyperpigmentation is more common in patients of color as darker colored skin has a greater tendency to hyperpigment following needle trauma. Hyperpigmentation is particularly seen in individuals with Fitzpatrick skin types IV, V, and VI.

Treatment for persistent post-inflammatory hyperpigmentation that occurs following HA dermal filler injection should include the application of topical bleaching agents such as topical hydroquinone (2–8 %) and Retin-A (tretinoin) [6]. In addition to bleaching agents, consistent daily sunscreen usage must be adhered to.

If the hyperpigmentation is resistant to this first line of treatment, chemical peels can be used [6]. If these treatments are unsuccessful, then laser treatment should be initiated. Laser choice will depend on skin type. IPL is effective in the treatment of Fitzpatrick skin types I–IV, while the Nd:YAG 1,064 nm laser has been effective for treating darker skin tones [6].

12.2.5 Dyschromia

Bluish discoloration under the skin is an adverse reaction seen particularly with hyaluronic acid fillers, usually the result of an improper injection technique whereby the filler in injected too superficially (or migrates superficially) [9]. Note that the color observed in patients has also been described as a grayish tint. The dermal hue change has been explained by the Tyndall effect, in which an optical phenomenon occurs as light is scattered as it passes through colloidal particles in solution. Since blue light, with a wavelength of 400 nm, scatters more readily than longer wavelengths, this is the predominant color seen when HA filler particles scatter light. While the Tyndall effect is the commonly accepted explanation for this particular dermal hue change, alternative explanations have been proposed to explain this discoloration [10]. While the Tyndall effect is known to be caused by a variety of hyaluronic acid derivatives, Belotero^{R}, a monophasic, highly cross-linked hyaluronic acid dermal filler, is reported not to cause the Tyndall effect [11].

This cause of dyschromia can usually be avoided if the product is injected at the correct dermal level. The more superficial the placement of the HA filler material, the longer the discoloration may last. To correct Tyndall effect discoloration due to HA, hyaluronidase is used. In addition, if necessary, surgical excision can be employed using a surgical scalpel (#11 blade) and then extruding the unwanted filler material [12]. The Nd:YAG 1064 nm laser has also been used successfully to treat this adverse reaction [13].

12.3 Nodules

12.3.1 Noninflammatory Nodules

Noninflammatory nodules are small palpable lumps that are oftentimes visible under the skin. These single, isolated lumps manifest at the injection site a few weeks following filler injection. Small nodule formation is an adverse effect mainly due to technical error and is commonly seen with superficial injection of HA fillers [14]. Nodules tend to occur around the mouth and eyes when dermal fillers are injected superficially. Thus, nodule formation is commonly due to improper superficial injection technique, as is dyschromia due to Tyndall effect discussed prior. In addition, noninflammatory nodules result from overcorrection, whereby an excessive amount of material has accumulated in the tissue. Furthermore, nodules may occur due to poor placement within highly mobile areas, such as the lips [14]. Nodules are less commonly seen with hyaluronic acid usage when compared to the particulate fillers calcium hydroxylapatite (CaHA) and poly-l-lactic acid (PLLA) [14].

Proper injection technique is paramount to minimize the formation of nodules. In the event of HA filler-induced nodules, hyaluronidase is the treatment of choice used to eradicate the subcutaneous nodule and to mitigate overcorrection. Hyaluronidase is an enzyme that dissolves hyaluronic acid in the skin and is employed to reverse the effects of HA filler injections. Before treating with hyaluronidase, a skin test must be performed to ensure there is no allergic response to hyaluronidase [15]. Anaphylaxis is a potential side effect of hyaluronidase, so it is important to ensure a negative allergic response test prior to administration of hyaluronidase.

12.3.2 Granuloma

In contrast to noninflammatory nodules, foreign body granulomas are inflammatory nodules caused by a nonallergic chronic inflammatory reaction. The resulting inflammatory lesion is predominantly composed of multinucleated giant cells and is caused by granulomatous inflammation after the aggregation of macrophages in response to large foreign bodies that cannot be phagocytosed by macrophages [16]. Filler-related foreign body granulomas typically occur 6–24 months after filler injection [17]. Foreign body granulomas are rare, with a reported incidence of foreign body granulomas after injection of hyaluronic acid of 0.02–0.4 %, with a peak estimated incidence of 1.0 % [18].

Clinically, foreign body granulomas caused by hyaluronic acid mainly appear as cystic granulomas and can be accompanied by edema and erythema. Development of a sterile abscess results from a process of encapsulation that prevents the absorption of the injected material into the surrounding tissues. Characteristic histological findings include palisaded granulomatous tissue composed primarily of giant cells and macrophages [19].

A differential diagnosis should be performed to distinguish granulomas from noninflammatory nodules. Filler-induced granulomas are differentiated from nodules in that the size of the granuloma becomes larger than the volume that was injected, and granulomas develop simultaneously at multiple sites of injection [18]. Since foreign body granulomas are not allergic reactions and are often triggered by a systemic bacterial infection, it is currently not possible to predict which patients are at risk for developing granulomas [18]. Left untreated, they may remain virtually unchanged for some years and then resolve spontaneously [18].

The primary treatment of foreign body granulomas caused by HA fillers is intralesional corticosteroid injections (betamethasone, prednisone, or triamcinolone) [20]. The local injection of corticosteroids disrupts the activities of fibroblasts, giant cells, and macrophages. Depending on the severity of the reaction, 5–10 mg/cc of corticosteroid should be used [9]. If necessary, repeat the corticosteroid injection 4–6 weeks later. For intralesional injections it is recommended that a 0.5 or 1.0 mL insulin syringe with a 30 gauge needle be used [17]. A smaller diameter syringe is advantageous as it allows the resistance of the granuloma to be sensed, which helps prevent corticosteroid-induced dermal atrophy [17]. As granulomas tend to spread into the surrounding tissue in a finger-like pattern, the preferred technique is to inject a small amount of drugs gradually, moving from the periphery to the central area [17]. To prevent recurrence, it is preferable to inject a high dose of triamcinolone mixed with lidocaine when performing intralesional steroid injections [17].

As an alternative treatment for granulomatous reactions, an injection containing bleomycin may work successfully [21]. In addition, 5-fluorouracil, an antimitotic agent, has been used in intralesional injections to treat granulomas [22]. Furthermore, granulomatous reactions to HA fillers have been treated with hyaluronidase [15].

Finally, systemic oral steroid therapy can be used for recurring foreign body granulomas. The use of oral prednisone at a starting dose of 30 mg/day and a maintenance dose and 60 mg/day can prevent the recurrence of granulomas [18]. Minocycline combined with oral or intralesional steroids is effective in treating widespread inflammatory granulomas [23].

The excision of foreign body granulomas is not a therapy of first choice because the complete removal of a granuloma is often not possible in many cases as granulomas are invasive and have non-confined borders with the surrounding tissue [24]. However, in the case of an obvious sterile abscess, an effective treatment is incision and drainage of the abscess [24].

12.4 Edema

12.4.1 Normal Edema

Edema is a common adverse reaction subsequent to an HA filler injection. Hyaluronic acid derivatives are particularly hydrophilic and can be associated with localized edema. Just as with erythema, edema is due to puncture trauma and associated inflammation. The swelling can be expected to persist for a similar duration as the erythema, sometimes slightly longer. Depending on the injection site, such as lip injection, swelling can be more profound and last longer, with an expected duration of 2–3 days.

Swelling is managed by applying cold compresses postinjection for 5–10 min to reduce inflammation. After the procedure, patients can be advised to use ice packs at home every few hours on the day of the injection. Caution must be emphasized to avoid prolonged ice pack usage to reduce the risk of cold injury to the skin. Furthermore, swelling can be reduced using a prudent injection technique that will minimize the number of skin punctures during the injection process, thus limiting trauma to the injected area. Such techniques include placing the filler with a serial injection, using the fanning technique, injecting at the preperiosteal level, or using linear tunneling with threading. Care must be taken when injecting HA fillers into regions such as the lower eyelids and lips where there is a greater likelihood of undesirable excessive visible edema. Additional volume may develop in these areas because of excessive water sequestration caused by hyaluronic acid derivatives [25]. Thus, conservative treatment should be undertaken when injecting the lower eyelids and lips.

12.4.2 Facial Angioedema

Facial angioedema is an adverse event that can occur following HA filler injection. Facial angioedema results from a hypersensitivity reaction, which is an allergic reaction mediated by T lymphocytes. This allergic reaction is thought to be related to protein contaminants present in the filler material. Hypersensitivity reactions related to dermal fillers are an infrequent complication. Immune-mediated angioedema is rare, with an estimated incidence of less than one to five in 10,000 [25].

Angioedema normally manifests within approximately 2 weeks posttreatment [25]. Angioedema is more commonly seen with superficially placed hyaluronic acid derivatives. A particular area of concern is the lip when injected superficially with HA fillers [25]. In the event of angioedema, the allergen (hyaluronic acid derivative) which is the inciting factor must be removed. This is accomplished by injecting hyaluronidase locally. If necessary, the symptoms of angioedema can be treated with oral prednisone [6].

12.5 Infection

12.5.1 Herpetic Reactivation

Herpes simplex virus reactivation has been reported following HA dermal filler injections, likely associated with the inherent skin irritation caused by injection. Common sites of reactivation are the perioral area, nasal mucosa, and mucosa of the hard palate [26]. These case reports are anecdotal and no definitive evidence-based data has implicated fillers in the causation of recurrent herpes infection [9]. However, for those patients with a history of cold sores, especially after prior filler injections, an antiherpes prophylaxis regimen may prove beneficial [27, 28]. Prophylactic treatment with valacyclovir should be initiated prior to injection to reduce herpetic reactivation, with a dosage of 500 mg twice daily for 3–5 days [6]. If a herpetic reactivation occurs in the absence of prophylactic treatment, then 2 g of valacyclovir twice daily for 1 day should be given to tamper the infectious outbreak.

12.6 Vascular Compromise

Vascular compromise following dermal filler injection is a rare but very serious potential adverse event, with an incidence estimate of 0.001 % for hyaluronic acid fillers [29]. Vascular compromise results from vessel injury, compression, or occlusion following dermal filler placement [14]. Oftentimes vascular compromise results when the intravascular injection of material into an artery occurs, causing an embolism that impedes blood flow [6]. Vascular injury can cause tissue necrosis and acute vision loss. To limit the risk of injection procedures, it is imperative that administrators of dermal fillers have a thorough understanding of facial anatomy. Vascular compromise is an emergent condition that requires prompt action to avoid catastrophic consequences.

12.6.1 Tissue Necrosis

Vascular-mediated events may result in skin necrosis following hyaluronic acid filler injection. Impending necrosis following filler injection is a major, early-onset complication that is likely the result of vascular injury, compression, or obstruction of the facial artery, angular artery, lateral nasal artery, supratrochlear artery, or their branches. During the injection procedure, filler material may inadvertently be injected into vessels and flow antegrade or retrograde through the vasculature, causing an occlusion leading to local or distal tissue necrosis [6]. In a review study of necrotic events following dermal filler injection, the most common injection site for necrosis was the nose (33.3 %), followed by the nasolabial fold (31.2 %) [30]. Necrosis can also occur due to vessel injury and compression secondary to the local edema caused by the hydrophilic properties of hyaluronic acid fillers [31].

The anatomic regions most susceptible to ischemic necrosis are the glabella and the nasolabial fold [14]. These are regions where the blood supply is poor or is predominantly dependent on a single arterial branch [32]. The glabella region is supplied by the supratrochlear and supraorbital arteries, terminal branches of the ophthalmic artery. Retinal artery occlusion can be caused by injections to the glabella, leading to vision impairment and complete vision loss [14]. The nasolabial fold is supplied by the angular artery. Alar necrosis has been reported following injection to the nasolabial fold, likely due to compression of the angular artery or its branches [25]. To prevent these serious adverse vascular events, extra caution must be taken when injecting into these areas. Aspiration prior to injection is recommended to help prevent accidental placement of the filling agent within a vessel. It is important to watch for the signs of vascular compromise which are severe pain (more than what is expected for a dermal filler injection) and an area of blanching [6]. If these symptoms occur, swift and aggressive treatment is necessary to prevent tissue necrosis.

In the event of hyaluronic acid-induced vascular compromise and impending necrosis, immediately discontinue the injection. Next, administer a cutaneous injection of hyaluronidase in the site of filler placement [33]. Then apply a 2 % nitroglycerin paste to the skin [31, 34]. Nitroglycerin paste has a vasodilatory effect on small-caliber arterioles, thus improving flow within the dermal vasculature. Apply the paste cyclically for 12 h on and 12 h off until clinical improvement. To further increase vasodilatation to the affected area, apply warm compresses and massage the area. In addition, aspirin 325 mg daily should be given to prevent clot formation [31]. Furthermore, methylprednisolone (Medrol dose pack) should be prescribed along with prophylactic antibiotic therapy such as levofloxacin [31]. Along with these measures, application of topical oxygen infusion cream (Dermacyte Oxygen Concentrate, Oxygen Biotherapeutics) twice daily has been reported effective [31]. Low molecular weight heparin has also been used in the management of patients with filler-induced vascular occlusion [35].

Factors that increase the possibility of vessel occlusion and resulting vascular compromise include high-pressure injections (anterograde flow more likely), large-volume bolus injections, a stationary rather than moving needle, and a deep plane of injection (larger vessels are found beneath the dermis in the subcutaneous fat) [6].

12.6.2 Vision Loss

The glabella is a high-risk anatomic location for ischemic necrosis. Accidental injection of the supratrochlear or supraorbital arteries in the glabellar region can cause a central retinal artery embolism that impedes blood flow to the retina resulting in visual impairment as a result of retrograde flow of the filler material into the central retinal artery. Precautions that can be taken to minimize the risk of central retinal artery embolism and iatrogenic blindness include aspirating before injection to detect accidental entry into a vessel; using needles and cannulas of small size as opposed to larger ones, and blunt flexible needles and microcannulas when possible; performing low-pressure injections with the release of the least amount of substance possible rather than bolus injections; and avoiding injection into traumatized tissue [36]. If visual impairment results after filler injection, an ophthalmologist should be consulted immediately (Figs. 12.1, 12.2, 12.3, 12.4, 12.5, 12.6, 12.7, 12.8, and 12.9).

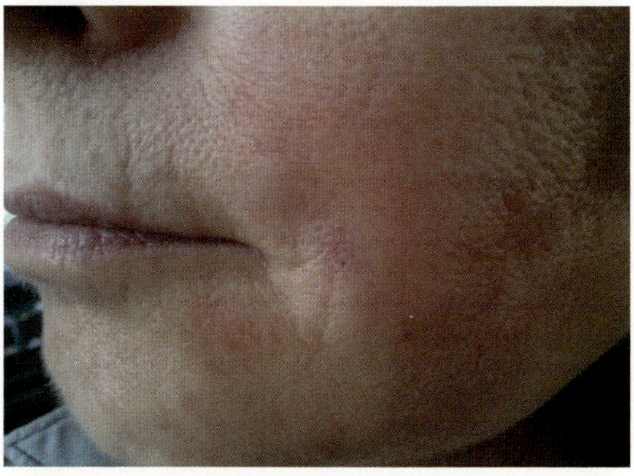

Fig. 12.1 Erythema and ecchymosis following injection of hyaluronic acid derivative filler

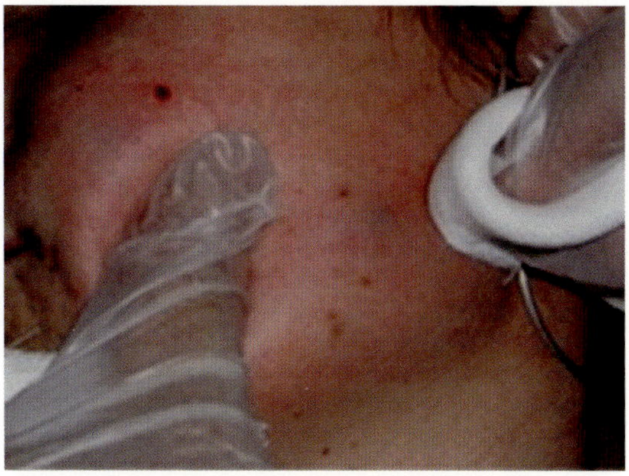

Fig. 12.2 Ecchymosis

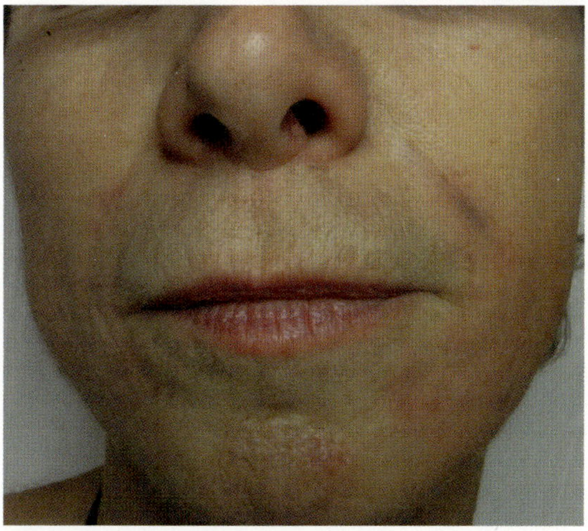

Fig. 12.3 Tyndall effect caused by superficial placement of hyaluronic acid derivative filler

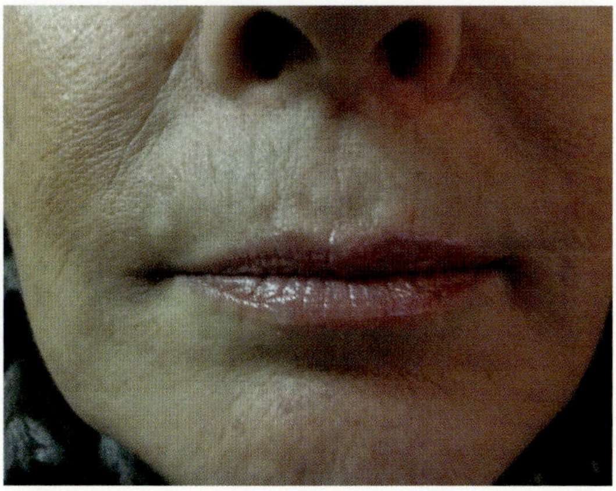

Fig. 12.4 Dermal nodule in perioral area

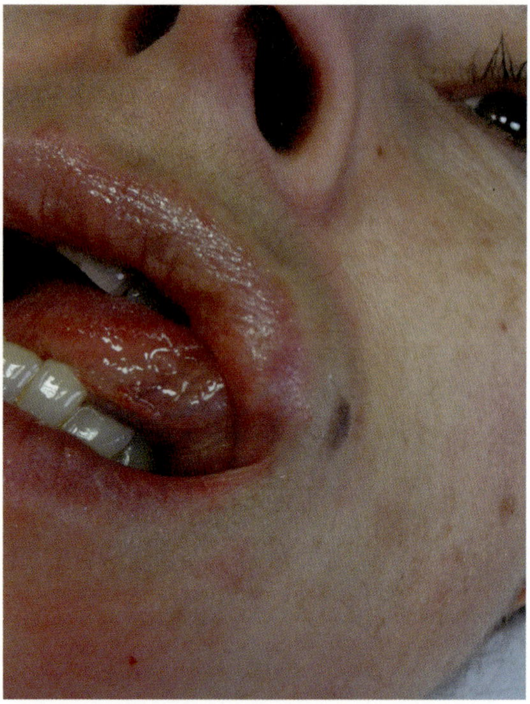

Fig. 12.5 Lip nodule associated with injection of hyaluronic acid

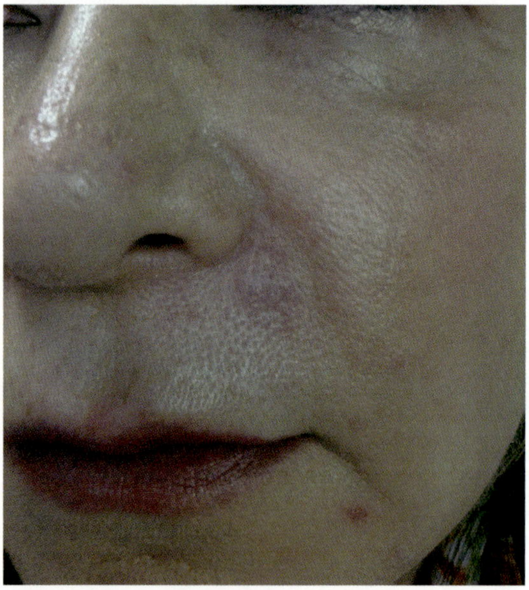

Fig. 12.6 Granuloma caused by hyaluronic acid injection

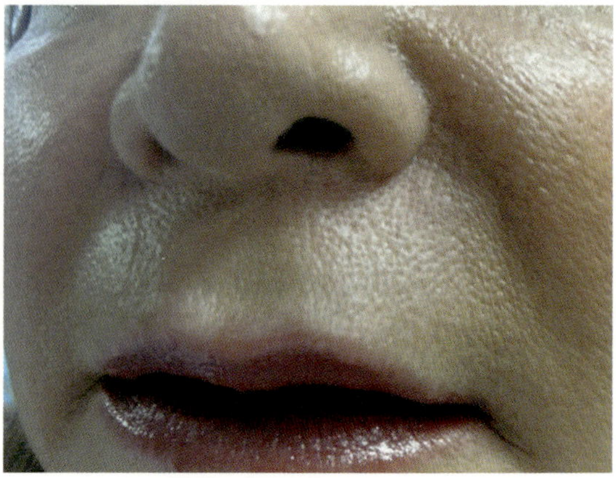

Fig. 12.7 Inflammatory foreign body granuloma

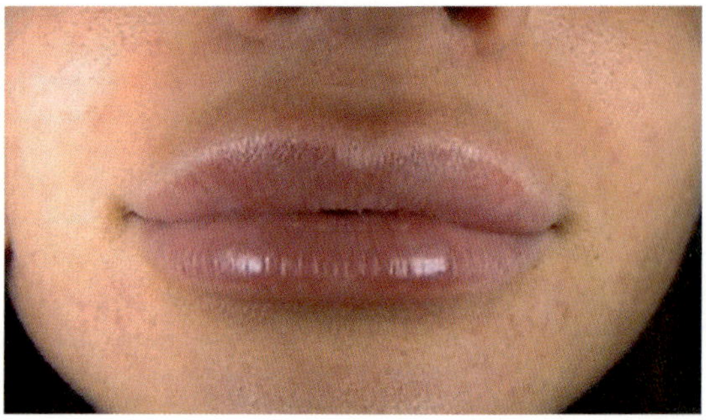

Fig. 12.8 Perioral and lip edema

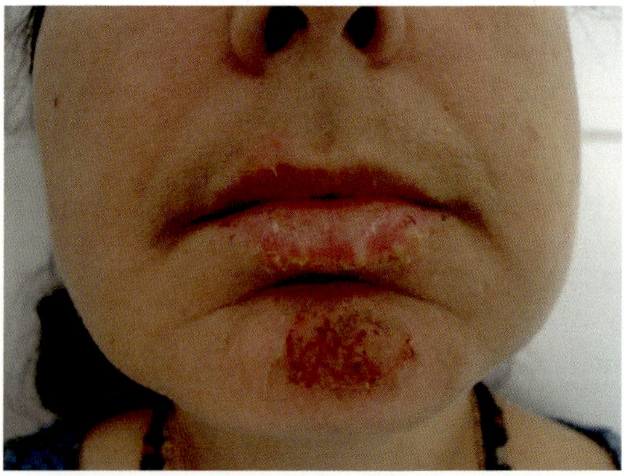

Fig. 12.9 Herpetic infection following injection with hyaluronic acid derivative filler

References

1. Funt D, Pavicic T (2013) Dermal fillers in aesthetics: an overview of adverse events and treatment approaches. Clin Cosmet Investig Dermatol 6:295–316. doi:10.2147/CCID.S50546
2. Gladstone HB, Cohen JL (2007) Adverse effects when injecting facial fillers. Semin Cutan Med Surg 26:34–39. doi:10.1016/j.sder.2006.12.008
3. Bailey SH, Cohen JL, Kenkel JM (2011) Etiology, prevention, and treatment of dermal filler complications. Aesthet Surg J 31:110–121. doi:10.1177/1090820X10391083
4. Cohen JL, Bhatia AC (2009) The role of topical vitamin K oxide gel in the resolution of postprocedural purpura. J Drugs Dermatol 8:1020–1024
5. Zeichner JA, Cohen JL (2012) Use of blunt tipped cannulas for soft tissue fillers. J Drugs Dermatol 11:70–72
6. Funt D, Pavicic T (2015) Dermal fillers in aesthetics. Plast Surg Nurs 35:13–32. doi:10.1097/PSN.0000000000000087
7. Lemperle G, Rullan PP, Gauthier-Hazan N (2006) Avoiding and treating dermal filler complications. Plast Reconstr Surg 118:92S–107S. doi:10.1097/01.prs.0000234672.69287.77
8. Sclafani AP, Fagien S (2009) Treatment of injectable soft tissue filler complications. Dermatol Surg 35:1672–1680. doi:10.1111/j.1524-4725.2009.01346.x
9. Lafaille P, Benedetto A (2010) Fillers: contraindications, side effects and precautions. J Cutan Aesthet Surg 3:16. doi:10.4103/0974-2077.63222
10. Rootman DB, Lin JL, Goldberg R (2014) Does the Tyndall effect describe the blue hue periodically observed in subdermal hyaluronic acid Gel placement? Ophthal Plast Reconstr Surg 30:524–527. doi:10.1097/IOP.0000000000000293
11. Pavicic T (2011) Efficacy and tolerability of a new monophasic, double-crosslinked hyaluronic acid filler for correction of deep lines and wrinkles. J Drugs Dermatol 10:134–139
12. Douse-Dean T, Jacob CI (2008) Fast and easy treatment for reduction of the Tyndall effect secondary to cosmetic use of hyaluronic acid. J Drugs Dermatol 7:281–283
13. Hirsch RJ, Narurkar V, Carruthers J (2006) Management of injected hyaluronic acid induced Tyndall effects. Lasers Surg Med 38:202–204. doi:10.1002/lsm.20283
14. Lolis M, Dunbar SW, Goldberg DJ et al (2015) Patient safety in procedural dermatology. J Am Acad Dermatol 73:15–24. doi:10.1016/j.jaad.2014.11.036
15. Brody HJ (2005) Use of hyaluronidase in the treatment of granulomatous hyaluronic acid reactions or unwanted hyaluronic acid misplacement. Dermatol Surg 31:893–897
16. Bentkover SH (2009) The biology of facial fillers. Facial Plast Surg 25:73–85. doi:10.1055/s-0029-1220646
17. Lee JM, Kim YJ (2015) Foreign body granulomas after the use of dermal fillers: pathophysiology, clinical appearance, histologic features, and treatment. Arch Plast Surg 42:232–239. doi:10.5999/aps.2015.42.2.232
18. Lemperle G, Gauthier-Hazan N, Wolters M et al (2009) Foreign body granulomas after all injectable dermal fillers: part 1. Possible causes. Plast Reconstr Surg 123:1842–1863. doi:10.1097/PRS.0b013e31818236d7
19. Micheels P (2001) Human anti-hyaluronic acid antibodies: is it possible? Dermatol Surg 27:185–191
20. Lupton JR, Alster TS (2000) Cutaneous hypersensitivity reaction to injectable hyaluronic acid gel. Dermatol Surg 26:135–137
21. Saray Y, Güleç AT (2005) Treatment of keloids and hypertrophic scars with dermojet injections of bleomycin: a preliminary study. Int J Dermatol 44:777–784. doi:10.1111/j.1365-4632.2005.02633.x
22. Conejo-Mir JS, Sanz Guirado S, Angel Munoz M (2006) Adverse granulomatous reaction to artecoll treated by intralesional 5-fluorouracil and triamcinolone injections. Dermatologic Surg 32:1079–1082. doi:10.1111/j.1524-4725.2006.32117.x
23. Lemperle G, Duffy D (2006) Treatment options for dermal filler complications. Aesthetic Surg J 26:356–364. doi:10.1016/j.asj.2006.04.002
24. Lemperle G, Gauthier-Hazan N (2009) Foreign body granulomas after all injectable dermal fillers: part 2. Treatment options. Plast Reconstr Surg 123:1864–1873. doi:10.1097/PRS.0b013e3181858f4f
25. Grunebaum LD, Bogdan Allemann I, Dayan S et al (2009) The risk of alar necrosis associated with dermal filler injection. Dermatol Surg 35(Suppl 2):1635–1640. doi:10.1111/j.1524-4725.2009.01342.x
26. Gazzola R, Pasini L, Cavallini M (2012) Herpes virus outbreaks after dermal hyaluronic acid filler injections. Aesthetic Surg J 32:770–772. doi:10.1177/1090820X12452293
27. Christensen L (2007) Normal and pathologic tissue reactions to soft tissue Gel fillers. Dermatol Surg 33:S168–S175.doi:10.1111/j.1524-4725.2007.33357.x
28. Christensen L, Breiting V, Janssen M et al (2005) Adverse reactions to injectable soft tissue permanent fillers. Aesthetic Plast Surg 29:34–48. doi:10.1007/s00266-004-0113-6
29. Narins RS, Jewell M, Rubin M et al (2006) Clinical conference: management of rare events following dermal fillers--focal necrosis and angry red bumps. Dermatol Surg 32:426–434. doi:10.1111/j.1524-4725.2006.32086.x
30. Ozturk CN, Li Y, Tung R et al (2013) Complications following injection of soft-tissue fillers. Aesthet Surg J 33:862–877. doi:10.1177/1090820X13493638
31. Dayan SH, Arkins JP, Mathison CC (2011) Management of impending necrosis associated with

soft tissue filler injections. J Drugs Dermatol 10:1007–1012
32. Bachmann F, Erdmann R, Hartmann V et al (2009) The spectrum of adverse reactions after treatment with injectable fillers in the glabellar region: results from the Injectable Filler Safety Study. Dermatol Surg 35(Suppl 2):1629–1634. doi:10.1111/j.1524-4725.2009.01341.x
33. Kim D-W, Yoon E-S, Ji Y-H et al (2011) Vascular complications of hyaluronic acid fillers and the role of hyaluronidase in management. J Plast Reconstr Aesthetic Surg 64:1590–1595. doi:10.1016/j.bjps.2011.07.013
34. Kleydman K, Cohen JL, Marmur E (2012) Nitroglycerin: a review of its use in the treatment of vascular occlusion after soft tissue augmentation. Dermatol Surg 38:1889–1897. doi:10.1111/dsu.12001
35. Kang MS, Park ES, Shin HS et al (2011) Skin necrosis of the nasal ala after injection of dermal fillers. Dermatol Surg37:375–380.doi:10.1111/j.1524-4725.2011.01891.x
36. Lazzeri D, Agostini T, Figus M et al (2012) Blindness following cosmetic injections of the face. Plast Reconstr Surg 129:995–1012. doi:10.1097/PRS.0b013e3182442363

13 Complications Associated with Botulinum Toxin Administration

Alexander Daoud and Martin Zaiac

13.1 Introduction

First approved in 1989 for the treatment of various neuromuscular disorders, it was not until 2002 that the US Food and Drug Administration (FDA) approved botulinum toxin for its first dermatologic application: enhanced cosmesis of glabellar lines. By 2004, the FDA extended its approval of onabotulinumtoxin A, or Botox®, to the treatment of primary axillary hyperhidrosis refractory to treatment with topical agents [1].

At present, injection of botulinum toxin is one of the most commonly employed modalities for facial cosmetic enhancement in the United States. According to the American Society for Aesthetic Plastic Surgery, neurotoxin injection was the most commonly performed nonsurgical cosmetic procedure in the year 2013: of the near 15.1 million cosmetic procedures performed that year, approximately 42 % (6.3 million) were botulinum toxin injections [1, 2]. Furthermore, axillary and palmar hyperhidrosis, which affect approximately 1–3 % of the population, had historically proven difficult to treat prior to the approval of botulinum toxin injections [3]. These findings demonstrate both the clinical and economic impact of botulinum toxin injections in modern medicine, as well as their essential place in the armamentarium of the procedural dermatologist.

While the side effect profile of botulinum toxin formulations is generally favorable, it is important for clinicians to be aware of the complications associated with their use. In this chapter, we explore several of these common and rare adverse effects, with focus on their typical clinical presentation and indications for management. Furthermore, we briefly discuss the emergence of patients presenting with complications following injection of illicit botulinum toxin-containing compounds in the hands of untrained or nonmedical personnel.

The original version of this chapter was revised.
An erratum to this chapter can be found at DOI 10.1007/978-3-319-23270-6_15

A. Daoud, BS
Greater Miami Skin and Laser Center, Florida International University, Miami, FL, USA

M. Zaiac, MD (✉)
Greater Miami Skin and Laser Center, Florida International University, Herbert Wertheim College of Medicine, Miami Beach, FL, USA
e-mail: drmartyz@aol.com

13.2 Botulinum Toxin: An Overview

Produced by *Clostridium botulinum*, an anaerobic, spore-forming bacterium, botulinum toxin is a zinc-containing neurotoxic enzyme that exerts its effect within the synaptic bouton of the neuromuscular junction. Through the hydrolysis of the proteins synaptobrevin (also referred to as

G. Fabbrocini et al. (eds.), *Nonsurgical Lip and Eye Rejuvenation Techniques*,
DOI 10.1007/978-3-319-23270-6_13

neuronal vesicle-associated membrane protein or VAMP), SNAP25, and syntaxin, botulinum toxin inhibits the release of the neurotransmitter acetylcholine, thereby inducing flaccid paralysis in affected muscles [4]. Pathologically, this effect is best demonstrated by the disease manifestations of botulism. Following ingestion and/or inhalation of clostridial spores, there is reactivation of the bacterial life cycle, with resultant production of massive amounts of botulinum toxin. Ultimately, the systemic release of this toxin load results in a clinical entity characterized by descending flaccid paralysis, respiratory arrest, and possibly death.

The strength of botulinum toxin is recorded as a measure of its paralytic activity in mouse species. The standard unit of injection, the unit (U), is described as the lethal dose for 50 % of mouse models, or LD_{50}, following intraperitoneal injection into the mouse abdomen. In humans, this dose has been estimated in the range of 3000 U [5, 6].

In 1989, the FDA approved the use of botulinum toxin type A – produced from the A subtype of *Clostridium botulinum* – as a local treatment for disorders ranging from blepharospasm and strabismus to various chronic facial spasm disorders. It was not until 2002 that botulinum toxin was approved for use in the management of moderate to severe glabellar lines. As with the neuromuscular disorders, its therapeutic effect was mediated by the induction of flaccid paralysis in muscles underlying the skin in which rhytides were present. Following the release of tension within these muscles – which typically takes about 14 days to reach maximal effect – there is a general flattening of the overlying skin that lasts for a period of 3–6 months. At this point, the recycling and regeneration of new neuromuscular junctions results in the reappearance of the original rhytides. Despite its effects at the neuromuscular junction, botulinum toxin does not appear to induce any reactive changes within myocytes themselves. In a clinicopathologic series performed on patients who received botulinum toxin type A injections in doses up to five times greater than those typically used for enhanced cosmesis, histologic examination did not demonstrate any chronic changes in muscle tissue, including scarring, fibrosis, or atrophy [7].

13.3 Botulinum Toxin Formulations in Clinical Practice

At present, there are three botulinum toxin-containing agents in use in the United States: onabotulinumtoxin A, abobotulinumtoxin A, and incobotulinumtoxin A. All three are derived from *Clostridium botulinum* serotype A, and they each possess their own unique clinical indications.

13.3.1 Onabotulinumtoxin A

Among the most commonly used botulinum toxin formulations on the market today, onabotulinumtoxin A – better known by its trade name, Botox® – is commonly used in cosmetic dermatology for the release of glabellar lines, hyperkinetic frontal lines, and lines of the lateral canthus ("crow's feet").

13.3.2 Abobotulinumtoxin A

Marketed under the trade name Dysport®, abobotulinumtoxin A is also used in clinical practice for facial cosmetic enhancement. However, it is important for clinicians to note that the relative potency of Dysport® does not equal that of Botox®: several studies have reported conversion factors ranging from 1:3 to 1:5 (Botox® vs. Dysport®). While the clinical effects after appropriate dosing are often similar between both agents, these findings are of considerable importance for clinicians or practices in which both toxin formulations are used interchangeably [6, 8].

13.3.3 Incobotulinumtoxin A

Appearing more recently on the market than the other two formulations, incobotulinumtoxin A (trade name Xeomin®) has been approved for the treatment of cervical dystonia, blepharospasm, and glabellar lines. In comparison to onabotulinumtoxin A- and abobotulinumtoxin A-containing formulations, Xeomin features a lower total load of protein; therefore, this compound was originally marketed as a potentially safer option due to a theoretically lower chance of inducing host immunologic response. However, one randomized, double-blinded trial comparing incobotulinumtoxin A to onabotulinumtoxin A failed to demonstrate any measurable difference in safety or neutralizing antibody generation between the two arms [1, 9].

13.4 Complications Associated with Botulinum Toxin Injection

A myriad of studies have demonstrated the relative safety and low side effect profile of all of the above formulations of botulinum toxin for local injection. Typically, the common adverse effects experienced by patients are similar for all three agents: these include reactions at the injection site (erythema, pruritus, hematoma formation, or transient rash), focal muscle weakness, and headache.

The risk of complication is related to both the dose administered and the site of injection. In one meta-analysis on onabotulinumtoxin A's safety in facial injections, it was found that adverse effect rates were much higher in onabotulinumtoxin A groups than in corresponding placebo groups. Additionally, injections at the glabella carried a higher risk of complications than those at the lateral canthus. Interestingly, the most commonly reported adverse effect at both sites was transient headache, which may have been an artifact of injection rather than the toxin itself [1, 10].

While many of the above complications are transient and do not require further management, several rare complications associated with botulinum toxin injection have been reported. As these manifestations may require further management, it is important for clinicians to be aware of their typical presentations.

13.4.1 Disorders of the Ocular Musculature: Lid Ptosis, Diplopia, and Strabismus

The muscles injected during the treatment of glabellar and canthal lines lie in close proximity to key muscles of ocular movement. Accordingly, it is important for dermatologists to remain cognizant of the ophthalmic complications associated with injection of botulinum toxin.

Among these, the most commonly reported ocular adverse effect is *eyelid ptosis*. Typically occurring following injection of the procerus muscle, the proposed mechanism of effect is lateral spreading of injected botulinum toxin through the orbital septum. This places the levator palpebrae superioris muscle of the upper eyelid at risk of paralysis, with resultant eyelid ptosis that can manifest within the first 2 weeks following injection. Clinically, this may manifest as a 1–2 mm depression of the affected eyelid; diagnosis may be made by comparison of the affected eye with the contralateral lid, as well as an obscuring of the upper pupillary rim on the affected side. While most cases of ptosis are mild and tend to resolve within 2–4 weeks of injection, evaluation with an ophthalmologist may be indicated in cases of severe visual impairment. In these cases, the use of mydriatic eyedrops may induce enough upper eyelid contraction to overcome the degree of induced ptosis. Clinicians may lower the risk of inducing eyelid ptosis by using concentrated solutions of botulinum toxin; this will diminish the migratory potential of a large bolus of dilute solution. Additionally, while gentle massage is advised to increase in-plane spread of toxin, avoidance of overaggressive horizontal massage will prevent the risk of toxin reaching the medial orbital septum [5].

Another ocular complication associated with botulinum toxin injection is *diplopia*, or the visual perception of a doubled image secondary to impaired extraocular muscle function. These findings are typically due to infiltration of injected botulinum toxin into the nearby extraocular muscles and often occur following injection of a large bolus of botulinum toxin or injection at the hands of untrained personnel. The typical patient presentation is an individual complaining of distorted vision 1–2 weeks following botulinum toxin injection. Paralysis of the lateral rectus muscle is among the most commonly reported complications; this may occur secondary to the regularity of the lateral canthus as a site of injection for enhanced cosmesis, as well as its close proximity to the lateral rectus muscle. However, excessive injection of the procerus or nasalis muscles may result in paralysis of the medial rectus muscle. Depending on the severity of diplopia, referral to ophthalmology may be warranted; however, clinicians can assure patients that this effect will reverse following regeneration of the neuromuscular junctions within the affected extraocular muscle.

A more severe manifestation of extraocular dysfunction is *strabismus*, or unilateral deviation of the affected globe secondary to the loss of function of an extraocular muscle. If suspected, urgent evaluation by an ophthalmologist is warranted: left unmanaged, these patients may experience long-term visual dysfunction. Ophthalmologists may choose to employ unilateral eye patching or treatment with visual glass prisms throughout the 3–6-month window until the effects of the toxin fade.

13.4.2 Complications Following Injection of Frontal Zone Musculature

Unlike eyelid ptosis, *eyebrow ptosis* typically occurs following injection of the frontalis muscle for the treatment of hyperkinetic frontal lines. This scenario often arises secondary to asymmetrical injection of botulinum toxin, injecting a large bolus of dilute solution in the frontal zone, or overaggressive horizontal massage following injection. It can be avoided by careful preparation in the pre-procedural window: using the smallest amount of concentrated injection material possible, as well as mapping injection sites prior to treatment, may prove helpful for clinicians.

Similar in presentation to eyebrow ptosis is an entity known as *pseudoptosis*. In the presence of redundant frontal skin – common in older patients or those with photodamaged skin – injection of the frontalis muscle may result in skin and subcutaneous tissue folding over the superior aspect of the brow. Counseling patients on the risks associated with frontalis muscle injection, as well as careful patient selection for injection at this site, may mitigate brow ptosis or pseudoptosis. Patients with this degree of cutaneous elasticity should also be advised that tissue edema following injection is common; this tends to resolve 24–48 h following injection.

Exaggerated elevation of the brow tail may occur in the setting of overaggressive treatment of the procerus muscle in comparison to the frontalis muscle. These patients will develop pronounced elevation of the lateral brow in comparison to the medial brow, which alters resting facial appearance and may interfere with normal expressions of emotion. This complication highlights a clinical pearl for botulinum toxin injection at any site: treatment of a muscle group (i.e., elevators) without concomitant injection of its antagonist group (i.e., depressors) may result in unfavorable cosmesis or distortion of resting facial structure.

13.4.3 Xerophthalmia (or "Dry Eye Syndrome")

The superolateral aspect of the bony orbit hosts the lacrimal fossa, within which the lacrimal gland – responsible for tear production – is situated. When botulinum toxin injectables were first approved for the treatment of lateral canthal lines, there was concern among clinicians that their proximity to this gland could pose a theoretical risk of iatrogenic *xerophthalmia*. However, contrasting reports on the effects of injection on tear production exist in the literature: while one prospective cohort of 26 crow's feet areas injected with botulinum toxin type A demonstrated no statistical difference in tear production (as measured by the Schirmer test), one recent study demonstrated that injection of the lateral canthal folds decreased both tear production and tear film stability. In this latter study, the severity of xerophthalmia was directly related to increasing patient age and increasing dose of botulinum toxin administered [11, 12].

Despite these findings, numerous case reports have described complications related to tear production following injection for lateral canthal rhytides. The generally accepted mechanisms of pathogenesis include direct toxin-mediated effects at the lacrimal gland, as well as orbicularis oculi muscle dysfunction secondary to toxin injection. The typical presentation mirrors that of xerophthalmia: patients report conjunctival injection, a sensation of "sand-like dryness," or general eye irritation in the weeks following botulinum toxin injection. If treatment is indicated, it is generally conservative and limited to management of the adverse effects experienced; patients can also be assured that their symptoms will improve as the effects of the botulinum toxin begin to fade. However, in cases where there is a significant degree of *lagophthalmos* (inability to close the eyelids), patients may present with ectropion (eversion of the lower eyelid), which places them at an increased risk of developing chronic keratitis and/or corneal ulcerations. In cases where this degree of lid dysfunction is suspected, urgent ophthalmological evaluation is warranted. Avoiding a large bolus of injection at the lateral canthus, as well as avoiding injection within one centimeter of the orbital ridge, can decrease the risk of botulinum toxin-induced xerophthalmia [13].

13.4.4 Complications Following Injections in the Middle to Lower Third of the Face

Many patients undergoing botulinum toxin injection may desire treatment of lower facial rhytides. Treatment in this region poses unique challenges to clinicians: a complex network of eleven separate levator and depressor muscles control lip movement, each with functions ranging from speech and eating to subtleties of facial expression. Therefore, improper treatment in this area may result in notable impairment for patients.

Ptosis of the upper lip may arise following either injection in the infraorbital or perizygomatic area, as well as injections affecting the superior aspect of the orbicularis oris muscle itself. For the former, resultant paralysis of lip elevators found in this region – including the zygomaticus major and minor, levator labii superioris, levator labii superioris alaeque nasi, and levator anguli oris – may result in asymmetrical drooping of the upper lip. Beyond the resultant unfavorable cosmesis, a significant degree of ptosis may interfere with normal speech, chewing, and facial expressions.

Conversely, *ptosis of the lower lip* may occur with improper injection in the region of the oral depressors, as well as migration of a large bolus into this region. Ptosis at this site may result in symmetric protrusion of the lower lip or may cause downward bowing of one oral commissure in comparison to the contralateral side. Beyond unfavorable cosmesis, such drooping of the oral commissures may interfere with drinking, eating, or speaking; if severe, patients may even experience spontaneous dripping of saliva from the affected oral commissure.

Injections in the area surrounding the chin are becoming increasingly popular, especially among male patients. Typically, patients present for evaluation of "chin furrowing," or corrugation of the skin overlying the chin secondary to a hypertrophic mentalis muscle, as well as relaxation of a "chin dimple," or midline cleft overlying the chin. Chemical denervation with botulinum toxin may ameliorate both conditions, but strict adherence to injections in the midline – as well as injecting at a safe distance from the orbicularis oris muscle – is essential at this site. Improper injection lateral to the mentalis may result in paralysis of the lower lip depressors, with resultant protrusion of the lower lip or ptosis. Additionally, migration upward toward the inferior rim of the orbicularis oris may prevent lip pursing and interfere with speech.

13.4.5 Complications Following Injection of the Neck and Platysma

Chemodenervation of the platysma provides patients with a safe, nonsurgical method for relaxing vertical neck bands. Nevertheless, the anterior neck is replete with neurovascular structures at potential risk of disruption, and the large surface area of the platysma often inclines clinicians to inject a high number of units of botulinum toxin (upwards of 100–200U in a single session has been reported by some). It is advised that clinicians with limited experience in injecting the platysma refer these patients to a procedural dermatologist or someone with greater familiarity of administering botulinum toxin in this area. Furthermore, the use of no greater than 50U of botulinum toxin is advised when injecting in the anterior neck.

Following injection of the platysma, common adverse effects include those related to the injection process itself: *pain, bruising, neck weakness, and generalized anterior neck discomfort* are often seen, and patients can be reassured that these symptoms will fade within several days. Although rarer, more alarming complications may arise following injections of the anterior neck: *impaired neck flexion, hoarseness of the voice, and dysphagia* have all been reported in the literature. Their mechanism is likely related to improperly injecting botulinum toxin deeply into the neck, as well as migration of a large bolus toward the deeper musculature of the neck. To avoid these potential complications, clinicians should use the smallest bolus of concentrated botulinum toxin possible, and they should inject superficially in a horizontal plane. Having the patient lie supine with their neck slightly flexed during injection may aid in preventing deep injections; additionally, following penetration of the skin, lifting up gently on the syringe to demonstrate a superficial location of the needle bevel will aid clinicians in administering the bolus in plane with the platysma.

13.4.6 Complications Following Botulinum Injection for Axillary and Palmar Hyperhidrosis

Historically, hyperhidrotic disorders have posed clinicians with challenges for long-term management. While the etiology of these disorders ranges from congenital to acquired, they are all believed to result from hyperactive eccrine gland secretion secondary to excessive stimulation by cholinergic sympathetic nerves. Accordingly, body areas with a high volume of eccrine glands – such as the axillae, palms, and soles – are typically the most affected in these disorders. In 2004, the FDA approved use of onabotulinumtoxin A for use in the treatment of focal axillary hyperhidrosis; however, both onabotulinumtoxin A and rimabotulinumtoxin B (trade name Myobloc; a botulinum toxin derived from *Clostridium botulinum* subtype B) are often used in clinical practice for the management of focal hyperhidrosis of other sites, including the soles and craniofacial areas [3, 17].

Several complications following treatment of hyperhidrotic disorders with botulinum toxin have been reported in the literature. Onabotulinumtoxin A has been reported as having a radial diffusion capacity of up to 1.5 cm within axillary skin, which makes mapping of the affected axillary skin essential prior to injection [17]. The starch-iodine test may prove useful in this regard, as it provides clinicians with a demonstrable area of involvement, which can serve as a guide for botulinum toxin injection [5]. If improperly injected, patients may experience *minimal clinical benefit and/or injection site reactions* secondary to material migration. Although exceedingly rare, there is one case report in the literature of a patient who developed *superficial thrombophlebitis* (Mondor's disease) of the anterior chest veins following injection of botulinum toxin subtype A for treatment of axillary hyperhidrosis [18]. These findings suggest that while botulinum toxin has proven effective in the management of axillary hyperhidrosis, administration of these injectable formulations is not without associated risks.

The most common complication of botulinum toxin injection for palmar hyperhidrosis is *hand weakness*: patients may report a general loss of dexterity that improves over the following 3–6 months. In order to minimize these complications, physicians should first map the injectable area and then administer the injection with a goal of distributing toxin within the superficial dermis [19].

13.4.7 Rare Complications Associated with Facial Injection of Botulinum Toxin: Superficial Temporal Artery Pseudoaneurysm and Systemic Manifestations of Local Injection

Many complications following injection of botulinum toxin have been reported in the literature, although cases are sporadic and often contested between studies. However, two of these entities are worth mentioning: superficial temporal artery pseudoaneurysm and systemic manifestations following local injection.

The superficial temporal artery is one of the terminal branches of the external carotid artery. In its course through the lateral face, the superficial temporal artery runs along the posterior aspect of the neck of the mandible and ascends ~1–2 cm anterior to the tragus in the preauricular area. It ultimately splits into two prominent branches – a frontal and parietal branch – both of which can be palpated for pulses or may be visible in certain individuals. Due to its superficial nature, the superficial temporal artery and its associated branches are at increased risk of trauma during both surgical and nonsurgical procedures of the lateral face. Several cases of *pseudoaneurysm* of these vessels have been reported several months following injection of botulinum toxin: patients typically presented with nontender, pulsating, or non-pulsating frontal and/or temporal masses that corresponded with the site of injection [14, 15]. Occasionally, a bruit may be auscultated over the mass. Diagnosis can be confirmed with Doppler ultrasound, which demonstrates blood flow with an outpouching mass in connection with the affected vessel. Depending on the resources available in one's community, prompt evaluation and management by a vascular surgeon or interventional radiologist is indicated.

Although exceedingly rare, physicians should remain aware of the risk of systemic manifestations following local botulinum toxin injection. There are several reports in the literature – occurring in both the cosmetic and facial spasmodic disorder settings – of injections of botulinum toxin inducing *systemic myasthenic crises* [5, 16]. These findings highlight the importance of obtaining a good clinical history – including a personal or family history of myasthenia gravis or other motor neuron disorders – in all patients before botulinum toxin injection. Accordingly, clinicians should not administer botulinum toxin injections to any patient with a personal history of disorders involving the motor neuron unit.

13.5 Non-dermatologist-Administered Botulinum Toxin Injection

At present, administration of botulinum toxin for enhanced facial cosmesis is still the most commonly performed nonsurgical cosmetic procedure in the United States. The relatively low adverse effect profile associated with these compounds, coupled with their ubiquitous presence throughout the United States, has resulted in the public's perception of botulinum toxin-containing agents as a safe option for enhancing facial cosmesis. Although it is strongly suggested that those seeking botulinum toxin injections consult a dermatologist, individuals throughout the medical community – including physicians of all specialties, nurse practitioners, physician's assistants, dentists, and registered nurses – can pursue certification and provide Botox® injection as a regular part of their clinical practice. These training certifications, which generally consist of a single-day 8-h course, may not provide healthcare providers with enough time to hone their skills in injecting botulinum toxin; accordingly, the risk of the above complications is generally higher.

Alarmingly, it has been the experience of the authors that there has been an increase in the number of patients presenting for evaluation of complications following botulinum toxin in the nonmedical/illicit setting and/or outside of the United States. Many of the complications experienced by these patients have been detailed above, with the most common complaint generally being asymmetrical facial tone and resultant lack of cosmetic benefit. The setting in which these injections are provided also poses a challenge for managing clinicians: as many of these patients are unaware of the type of substance they received as an injection, there is a chance that they have received non-FDA-approved botulinum toxin agents and/or formulations containing many different compounds. Clues to the latter include induration or prolonged erythema at the injection site, local tissue necrosis, and/or soft tissue hardening as a result of chronic inflammation and fibrosis.

While granuloma formation following injection of onabotulinumtoxin A has been reported in the literature, these cases are the subject of controversy among clinicians; in the presence of a positive history of facial injections performed outside of the medical setting in the United States, the presence of a granuloma strongly suggests injection of a material other than botulinum toxin [20–22]. In cases of unclear etiology, pathologic assessment of a biopsy specimen can reveal the presence of foreign bodies, including silicone globules or other foreign injection materials. While medical management with immunosuppressive agents such as cyclosporine or oral steroids may prove useful in

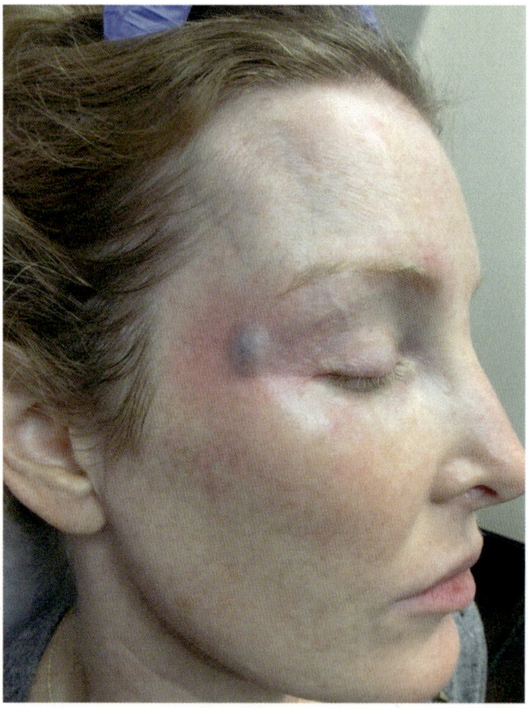

Fig. 13.1 Hematoma formation following botulinum toxin injection, right lateral canthus. Injection site hematomas occur most commonly at sites with rich vasculature and thin overlying skin, such as the periorbital and perioral regions, although they may occur anywhere. As with superficial hematomas at other body sites, patients can be assured that the area of raised and discolored skin will fade over the course of 1–2 weeks. Warm and cool compresses may be used if patients complain of pain at these sites

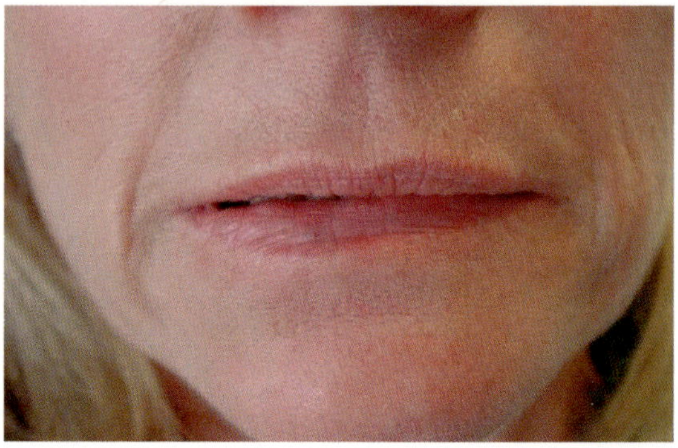

Fig. 13.2 Lower face asymmetry: ptosis of the left lower lip. Improper injection in the region of the oral depressors, as well as migration of a large bolus into this region, may result in lower lip ptosis, as is seen in this patient. There is notable downward bowing of the right oral commissure in comparison to the left, with persistent opening of the oral cavity that may interfere with drinking, eating, or speaking. If severe, patients may even experience spontaneous loss of saliva on the affected side

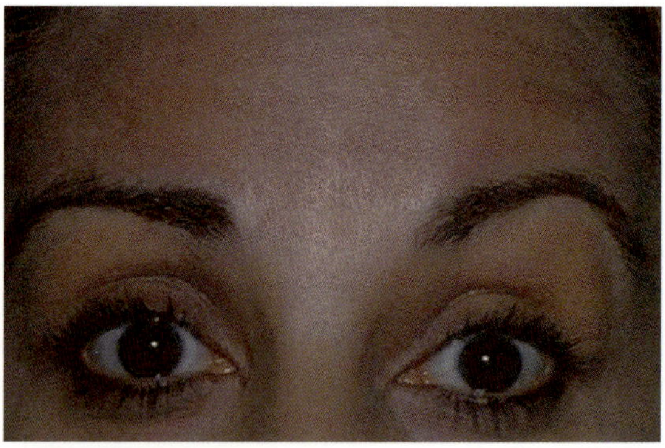

Fig. 13.3 Upper face asymmetry: exaggerated elevation of the brow tail. As compared to the right brow, there is marked elevation of the left brow tail in the resting face of this patient. Such a complication commonly occurs following inadequate treatment of the ipsilateral portion of the frontalis muscle in the setting of an adequately treated procerus muscle. The unopposed levator action of the frontalis, coupled with relaxation of the medial brow following treatment of the procerus muscle, results in an exaggerated elevation of the brow, most notable at the brow tail

uncomplicated cases of granuloma formation, surgical management – including excision and local tissue debridement – is most likely warranted.

Conclusion

When administered by trained medical professionals, botulinum toxin injections provide patients with a safe, nonsurgical method for both cosmetic and medical conditions alike. Nevertheless, it is important for clinicians to remain mindful that these agents are not inert: as a neurotoxic compound, botulinum toxin can pose significant morbidity to patients when injected improperly. Awareness of both the common and rare side effects associated with botulinum toxin injection encourages best practice standards whenever botulinum toxin is injected and also facilitates prompt evaluation and management in the event that a patient experiences any of these described complications (Figs. 13.1, 13.2, and 13.3).

References

1. Lolis M, Dunbar SW, Goldberg DJ, Hansen TJ, MacFarlane DF (2015) Patient safety in procedural dermatology: part II. Safety related to cosmetic procedures. J Am Acad Dermatol 73(1):15–24
2. Sorensen EP, Urman C (2015) Cosmetic complications: rare and serious events following botulinum toxin and soft tissue filler administration. J Drugs Dermatol 14(5):486–491
3. Strutton DR, Kowalski JW, Glaser DA, Stang PE (2004) US prevalence of hyperhidrosis and impact on individuals with axillary hyperhidrosis: results from a national survey. J Am Acad Dermatol 51(2):241–248
4. Nigam PK, Nigam A (2010) Botulinum toxin. Indian J Dermatol 55(1):8–14. doi:10.4103/0019-5154.60343
5. Klein AW (2004) Contraindications and complications with the use of botulinum toxin. Clin Dermatol 22(1):66–75
6. Wollina U, Konrad H (2005) Managing adverse events associated with botulinum toxin type A: a focus on cosmetic procedures. Am J Clin Dermatol 6(3):141–150, Review
7. Borodic GE, Ferrante R, Pearce LB, Smith K (1994) Histologic assessment of dose-related diffusion and muscle fiber response after therapeutic botulinum A toxin injections. Mov Disord 9(1):31–39
8. Sampaio C, Costa J, Ferreira JJ (2004) Clinical comparability of marketed formulations of botulinum toxin. Mov Disord 19(Suppl 8):S129–S136
9. Lee JH, Park JH, Lee SK, Han KH, Kim SD, Yoon CS, Park JY, Lee JH, Yang JM, Lee JH (2014) Efficacy and safety of incobotulinum toxin A in periocular rhytides and masseteric hypertrophy: side-by-side comparison with onabotulinum toxin A. J Dermatolog Treat 25(4):326–330
10. Brin MF, Boodhoo TI, Pogoda JM, James LM, Demos G, Terashima Y, Gu J, Eadie N, Bowen BL (2009) Safety and tolerability of onabotulinumtoxinA in the treatment of facial lines: a meta-analysis of individual patient data from global clinical registration studies in 1678 participants. J Am Acad Dermatol 61(6):961–70.e1-11
11. Arat YO, Yen MT (2007) Effect of botulinum toxin type a on tear production after treatment of lateral canthal rhytids. Ophthal Plast Reconstr Surg 23(1):22–24
12. Ho MC, Hsu WC, Hsieh YT (2014) Botulinum toxin type a injection for lateral canthal rhytids: effect on tear film stability and tear production. JAMA Ophthalmol 132(3):332–337
13. Ozgur O, Murariu D, Parsa AA, Parsa FD (2012) Dry eye syndrome due to Botulinum Toxin type-A injection: guideline for prevention. Hawai'i J Med Public Health 71(5):120–123
14. Prado A, Fuentes P, Guerra C, Leniz P, Wisnia P (2007) Pseudoaneurysm of the frontal branch of the superficial temporal artery: an unusual complication after the injection of botox. Plast Reconstr Surg 119(7):2334–2335
15. Skaf GS, Domloj NT, Salameh JA, Atiyeh B (2012) Pseudoaneurysm of the superficial temporal artery: a complication of botulinum toxin injection. Aesthetic Plast Surg 36(4):982–985
16. Borodic G (1998) Myasthenic crisis after botulinum toxin. Lancet 352(9143):1832
17. Glaser DA, Galperin TA (2014) Local procedural approaches for axillary hyperhidrosis. Dermatol Clin 32(4):533–540
18. Pisani LR, Bramanti P, Calabro RS (2015) A case of thrombosis of subcutaneous anterior chest veins (Mondor's disease) as an unusual complication of botulinum type A injection. Blood Coagul Fibrinolysis 26
19. Lehman JS (2011) Writer's block: "texting" impairment as a complication of botulinum toxin type A therapy for palmar hyperhidrosis. Arch Dermatol 147(6):752
20. Yun WJ, Kim JK, Kim BW, Lee SK, Kim YJ, Lee MW, Chang SE (2013) The first documented case of true botulinum toxin granuloma. J Cosmet Laser Ther 15(6):345–347
21. Pontes HA, Pontes FS, de Oliveira GF, de Almeida HA, Guimarães DM, Cavallero FC (2012) Uncommon foreign body reaction caused by botulinum toxin. J Craniofac Surg 23(4):e303–e305
22. Styperek A, Bayers S, Beer M, Beer K (2013) Nonmedical-grade injections of permanent fillers: medical and medicolegal considerations. J Clin Aesthetic Dermatol 6(4):22–29

Complications of Fractional Lasers (Ablative and Non-ablative)

14

Norma Cameli and Maria Mariano

14.1 Introduction

Targeted and innovative techniques and protocols are increasingly used in noninvasive eye and lip rejuvenation with the aim to obtain the best results and reduce side effects.

Understanding the anatomy of the eyelids, lips, and surrounding structures is important to achieve the best results and avoid potential complications.

Palpebral area is very delicate consisting of three layers: cutaneous, muscular, and fibrous layers.

In particular, palpebral skin is thinner compared to other skin districts; its hydrolipidic film and skin barrier function are weaker. The dermis is poorly represented and less rich in collagen and elastic fibers. The hypodermis is almost absent and blood and lymphatic circulations are slow.

Eyelid skin shows vascular fragility and increased vulnerability to actinic damage. The use of lasers to treat the eyelids is often limited by longer postoperative wounding, erythema, and the potential risk for hypopigmentation and ectropion.

The upper lip extends from the base of the nose superiorly to the nasolabial folds laterally and to the free edge of the vermilion border inferiorly. The lower lip extends from the superior free vermilion edge superiorly, to the commissures laterally, and to the mandible inferiorly. From superficial to deep, the layers of the upper and lower lips include the epidermis, subcutaneous tissue, orbicularis oris muscle fibers, and mucosa.

Fractionated laser technology has allowed physicians to minimize downtime and complications increasing the number of treatments with lower rate of complications than non-fractionated laser treatment [1]. While ablative fractional devices allow for quicker recovery than traditional fully ablative devices, when compared with their non-ablative counterparts, downtime can be considerably longer, in average 5–7 days.

Unfortunately, adverse effects can still occur even with the best technology and physician care.

Non-ablative fractional lasers (NAFL) are more gentle than the ablative and require a moderate amount of downtime as they induce limited tissue damage and melanocyte stimulation. In general, NAFR has fewer complications than traditional ablative lasers. Most complications can be easily managed and are self-limited. With regard to any side effect, early identification and treatment will improve outcome.

Ablative fractionated lasers (AFL) reduce the tissue trauma decreasing downtime while

N. Cameli (✉) • M. Mariano
Department of Dermatology, San Gallicano Dermatological Institute – IRCCS,
Via Elio Chianesi 53, Rome 00144, Italy
e-mail: cameli@ifo.it; mariano@ifo.it

G. Fabbrocini et al. (eds.), *Nonsurgical Lip and Eye Rejuvenation Techniques*,
DOI 10.1007/978-3-319-23270-6_14

retaining resurfacing action. These lasers are significantly safer than their non-fractionated counterpart, but they still maintain high risk of potential damage and complications.

Complication prevention, detection, and treatment are an important part of a physician's ability to provide the best results when treating a patient with fractionated laser.

14.2 Technology

Fractional lasers perform a pixelated pattern photothermolysis. This technology makes it possible to obtain microareas of thermal damage surrounded by healthy tissue. These microcolumns of damage stimulate the healing and skin restructuring processes with the production of new collagen and elastin, similar to those achieved with massive treatment of the entire surface, but instead limited to dots of a diameter of 70–150 μm separated by bridges of untouched skin. It has been estimated that the thermal damage induced by these microcolumns reaches a depth of between 300 and 400 μm in the dermis [2–4]. Histological studies by Hantash et al. [5] demonstrated that areas of epidermal and dermal necrosis are visible in the skin immediately after treatment that rapidly heal within 24 h, showing keratinocyte migration and elimination of the necrotic epidermal columns through exfoliation of the stratum corneum. Changes in cell morphology have also been observed in the deeper portions of the "columns." Specifically, stationary cuboidal phenotypes and even spindle cell migration are visible. These cells are considered to be responsible for the rapid healing and reepithelialization phenomena after fractional laser treatment.

Fractional CO_2 laser combines the "concept" of fractional photothermolysis with an ablative wavelength of 10,600 mm, successfully treating photoaging, acne scars, and skin flabbiness with minimized postoperative risks and discomfort.

Fractional CO_2 laser treatment does not require general anesthesia; however, a cooling system is implemented and topical anesthetics can be applied beforehand. Erythema may appear and lasts 5–7 days only, with only minimal risks of post-inflammatory hyperpigmentation and superinfections. The treatment requires more sessions than normal CO_2 laser treatment and the results are slower; however, patients prefer fractional laser treatment since it ensures faster healing times without any restrictions to their daily activities.

The non-ablative fractionated lasers combine the gentle and safe aspects of fractionated and non-ablative technologies aiming to improve texture, mild to moderate wrinkles, and acne scarring.

In general, NAFR has fewer complications than traditional ablative lasers. Most complications can be easily managed and are self-limited. As with any side effect, early identification and treatment will improve outcome.

14.3 Minor and Short-Term Complications

14.3.1 Erythema

Despite concerns for erythema, it should be kept in mind that erythema is the clinical end point of fractional resurfacing and is an expected, transient side effect.

In the case of non-ablative or ablative fractional resurfacing, redness may persist for 3–7 days.

For NAFR, prolonged erythema is defined as posttreatment redness that persists longer than 4 days. It has been reported in less than 1 % of patients treated with NAFR.

Ablative fractional resurfacing erythema has a longer duration. Usually post-resurfacing erythema fades gradually over time.

Prolonged erythema (Fig. 14.1a, b) can be caused by inappropriate laser settings, infections, and contact dermatitis.

Patients can be started on a topical steroid (hydrocortisone 2 %) to reduce inflammation.

Transient erythema after non-resurfacing procedures could be covered with cover-up makeup.

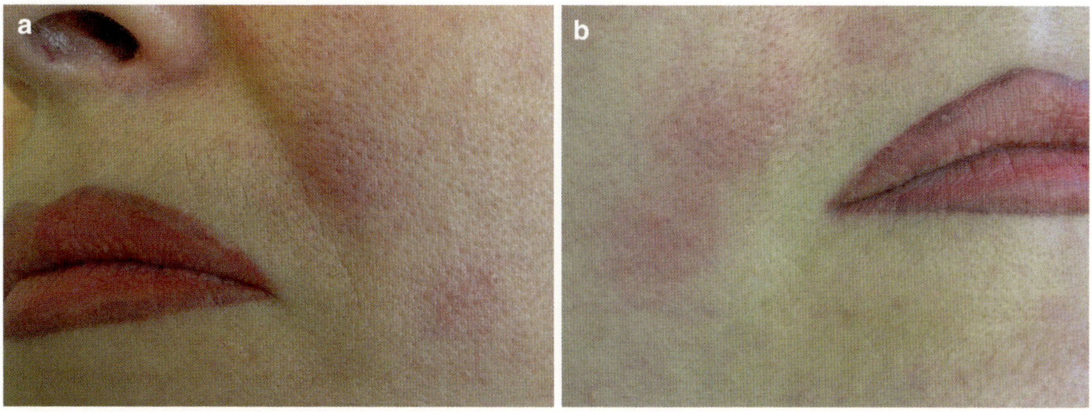

Fig. 14.1 (**a**) Persistent erythema 1 month after perioral AFR. (**b**) Persistent erythema 1 month after perioral AFR

14.3.2 Edema

After laser resurfacing mild edema could appear together with erythema, remissioning in automatic. Especially at the level of the eyelids, laser treatment produces marked edema, which can be notable for several days. The edema of the eyelids after laser resurfacing can get worse for 1–2 days after the procedure before it starts to reduce because it tends to congregate at eyelid levels. When needed edema can be treated with oral corticosteroids such as methylprednisolone in a brief course of 5 days (60 mg daily).

14.3.3 Urticaria

Immediate posttreatment urticaria is an expected consequence of fractionated laser skin resurfacing that usually resolves within 3–4 days. Cold-induced urticaria has been described after fractional carbon dioxide laser resurfacing of the face associated with cooling systems used during the procedure [6]. A complete medical history before starting treatment could be useful for prevention.

14.3.4 Ecchymoses

Petechiae or purpura can occur immediately or days after treatment and can take 1–2 weeks to resolve [7]. Postprocedure bruising can be minimized by avoidance of anticoagulants and other medication that may predispose (e.g., aspirin, vitamin E, ginkgo biloba, etc.).

Intense bruising could resolve leaving post-inflammatory hyperpigmentation, especially in photodamaged individuals and darker skin types.

14.3.5 Crusting and Erosions

Focal crusting and erosions may frequently occur during a non-ablative procedure. Erosions or crusts lasting more than 2–3 days should be treated with a brief course of topical steroids. Persisting lesions should indicate other causes, such as infections, inappropriate laser settings, or picking behavior.

After AFR larger areas of disepithelialization could be commonly seen and resolve in about a week.

Treatment of the eyelids with AFR in particular can cause erythema, edema, and focal erosions, visible for 3–5 days after treatment.

Posttreatment with abundant emollients can accelerate healing.

14.3.6 Blistering

Widespread small vesicles may be a reactive phenomenon after fractional laser treatments, especially

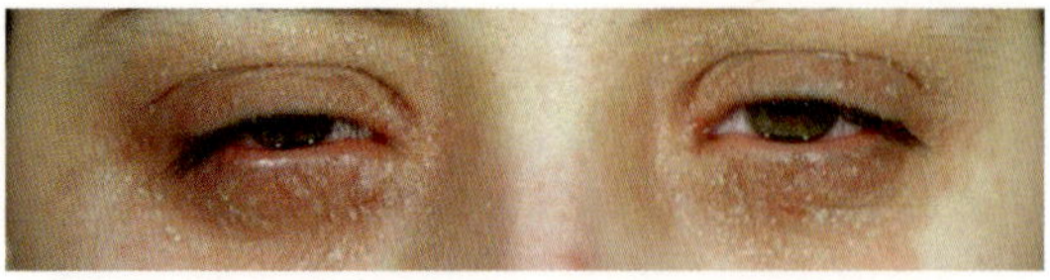

Fig. 14.2 Allergic contact dermatitis of the eyelids after non ablative fractional laser due to methylisothiazolinone contained in a postreatment cream

in eyelid regions. These lesions resolve within a day or two helped by topical corticosteroid application.

14.3.7 Contact Dermatitis

The thin skin of the eyelids is particularly sensitive to irritants and allergens and is thus prone to develop contact dermatitis due to the irritant and/or allergic potential of pre- and posttreatment topical agents (Fig. 14.2). It is recognized that a wide variety of creams, ointments, cleansers, and other skin care products may cause contact dermatitis after laser resurfacing [8]. Contact with the same trigger may not lead to a rash on other areas of the skin.

Gentle skin care and topical corticosteroids are recommended if needed.

14.4 Moderate and Medium-Term Complications

14.4.1 Acneiform Eruptions and Milia

Acneiform eruption incidence has been reduced by fractional technology comparing to traditional laser resurfacing. After NAFR treatments the rates of acneiform eruptions range from 2 to 10 %; milia can occur in up to 19 % of treated patients [9].

AFR treatments also show lower risk of developing acneiform eruptions or acne exacerbation and milia, which may be due to occlusive moisturizer application in the postoperative period [10]. Acne and milia often resolve without additional intervention as the healing processes. Nonocclusive and noncomedogenic moisturizers may help in reducing their incidence.

14.4.2 Infection

The most common infection after fractional laser skin resurfacing is related to the herpes simplex virus (HSV), with reported rates ranging from 0.3 to 2.76% [9, 11]. The incidence of bacterial infection after NAFR appears extremely low with 0.1 % of all treated cases documented to develop impetigo [9]. The infection rates with traditional ablative laser resurfacing were much higher, with 2–7 % of cases developing HSV reactivation [12].

Herpes simplex reactivation could be very common without prophylaxis. Patients may not present with classic herpetiform vesicopustules, but instead may demonstrate only superficial erosions that develop during the first week after treatment [11, 13].

To minimize the risk of HSV reactivation with fractional resurfacing, antiviral prophylaxis should be administered when a prior history of facial HSV is documented or if full-face ablative laser procedures are performed. Prophylactic therapy is important even in those without a history of herpetic infections. All patients should be placed on antiviral prophylaxis, starting the day before the procedure in those without history of herpetic infections and 3 days before in those with history of herpetic infection. Antiviral therapy should be continued for a total of 10 days.

The most common causes of skin infections after fractional resurfacing include *Staphylococcus aureus* (Fig. 14.3), *Pseudomonas*, *Klebsiella*, and *Enterobacter*. Persistent pruritus and prolonged erythema may be associated with candidiasis. Atypical mycobacterial infection has also been reported [14–17]. For this reason, many practitioners prescribe oral antibiotics and antivirals before starting the procedure continuing until skin reepithelialization is almost complete. Even if prophylactic antibiotics and antivirals have been used, in suspicion of skin infection, microbiologic culture testing should be conducted to identify the organism and its sensitivity to treatment.

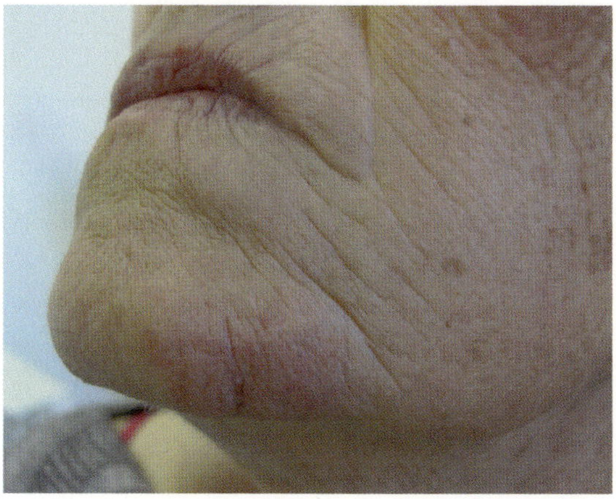

Fig. 14.3 *Staphylococcus aureus* skin infection after AFR

14.4.3 Dyspigmentation

Hyperpigmentation is one of the more common side effects of cutaneous laser resurfacing and may be expected to some degree in all patients with darker skin tones. Post-inflammatory hyperpigmentation (Fig. 14.4) is much less frequent with fractional laser skin resurfacing than with their non-fractional counterparts. Even though it is observed in 1–32% of patients [9, 13] depending on the device used, setting parameters, and Fitzpatrick skin phototype. The reaction is transient, but its resolution may be hastened with the postoperative use of a variety of topical agents, including hydroquinone and retinoic, azelaic, and glycolic acid. Darker skin phototypes (Fitzpatrick III–VI) have higher susceptibility for developing hyperpigmentation after AFR. NAFR are associated with very low rates of post-inflammatory hyperpigmentation, darker skin phototypes being more prone to develop it.

In general fractional laser treatments of darker skin should use higher fluencies, lower densities, and longer intervals between treatments.

Regular sunscreen use is also important during the healing process to prevent further skin darkening.

Hypopigmentation is not a common complication of AFR.

Postoperative hypopigmentation is often not observed for several months and is particularly difficult because of its tendency to be not responsive to treatment [18].

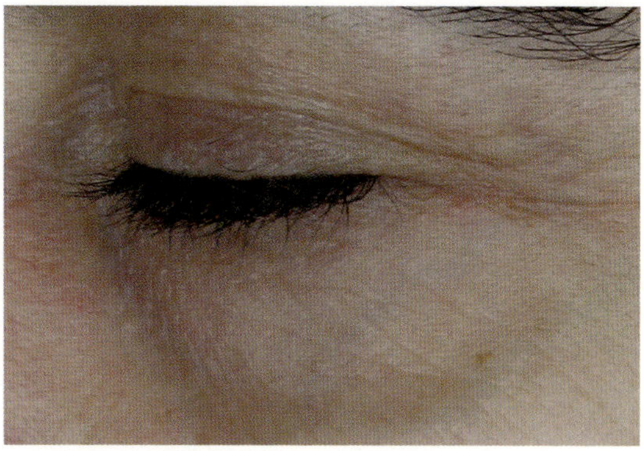

Fig. 14.4 Post-inflammatory hyperpigmentation after AFL of the eyelids (1)

14.4.4 Textural Defects

When the treatment is performed on a single part of an anatomic site, like perioral regions or eyelids, it can result in smoothness consistency of the treated side versus the surrounding skin. In general it resolves spontaneously.

14.4.5 Delayed Reepithelialization

Delayed reepithelialization may occur following the application of resurfacing lasers. When it doesn't occur in about a week, other causes should be investigated; the most frequent is infection.

It is extremely important to manage this uncommon complication because the longer the skin repairs, the higher is the risk of scarring.

14.5 Severe and Long-Term Complications

14.5.1 Scarring

Although the risk of scarring has been significantly reduced with the newer pulsed systems (compared with the continuous wave lasers), inadvertent pulse stacking or scan overlapping, poor technique, as well as incomplete removal of desiccated tissue between laser passes can cause excessive thermal injury that could increase the development of fibrosis [19, 20].

The most common cause of scarring is postoperative infection. Focal areas of bright erythema, with pruritus, may signal impending scar formation. Ultrapotent topical corticosteroid preparations should be applied to decrease the inflammatory response. A pulsed dye laser (PDL) can also be used to improve the appearance and symptoms of laser-induced burn scars.

The periorbital and mandibular regions are scar-prone anatomic locations that require more conservative treatment protocols.

14.5.2 Ocular Injury

Eye damages caused by laser procedures are not very common complications secondary to the use of inappropriate safety measures. Ocular injuries reported during laser use include coloboma and corneal, vitreous, and retinal damage. Before the laser is turned on, in ready mode patients' eyes should be closed or covered with opaque goggles or eye shields. The operator and other personnel in the room should wear filter glasses that selectively exclude wavelengths emitted by the laser. Laser-protective eyewear is a well-recognized precaution and includes wraparound glasses and goggles, which are rated by optical density (OD) at various wavelengths.

14.5.3 Ectropion

Ectropion of the lower eyelid after periorbital fractional laser is rarely seen. It is more frequent in patients who have had previous lower blepharoplasty or other surgical manipulations of the periorbital region. Preoperative clinical evaluation is important to determine eyelid skin laxity and elasticity. Lower fluences and fewer laser passes should be performed in the periorbital area to decrease the risk of lid eversion. When ectropion occurs, it usually requires surgical correction.

14.5.4 Koebnerization

Laser-induced trauma may initiate a koebnerizing dermatosis, including diseases such as vitiligo and psoriasis. Eruptive keratoacanthomas have been reported, most likely secondary to koebnerization [21].

14.6 Prevention

Laser safety includes the use of protective eyewear and eye shields, laser signage, control of surgical smoke, tissue splatter and plume, and attention to non-beam and beam hazards.

Treatment setting regulation is important to prevent side effects. In particular, when treating eyelids laser settings should be lower than those for non-eyelid skin because of the thinness of eyelid skin. In particular treatment with fractional lasers should be approached with lower fluence, lower density, and shorter pulse duration settings.

Appropriate precooling, cooling during the procedure, and postcooling should also be considered to provide an extra measure of epidermal protection. In addition, a detailed disclosure of potential side effects protects not only the patient but also the provider.

In preventing fractional laser complications while treating eyelids, the lips and perioral region are also important to perform correct pre- and post-laser skin care. Regular use of sunscreens and avoidance of tanning should be started on a preoperative regimen about a full month in advance continuing sun avoidance as skin care practice after the resurfacing.

As regards to post-laser treatment skin care, for patients undergoing NAFL resurfacing, the care is minimal. The use of a mild, fragrance-free cleanser and moisturizer could be recommended resuming regular skin care regimen after about 1 week. For patients undergoing AFL, some surgeons recommend cleaning process with only tap water and gentle gauze followed by application of a light lubricating ointment. Others add the use of local antibiotics and/or antifungal. When the resurface is complete (between days 4 and 6 after the procedure) but redness has not faded, fragrance-free cleanser and moisturizer should be used. Makeup is allowed once resurfacing is complete preferring mineral makeup.

References

1. Cameli N, Mariano M, Serio M, Ardigò M (2014) Preliminary comparison of fractional laser with fractional laser plus radiofrequency for the treatment of acne scars and photoaging. Dermatol Surg 40(5):553–561
2. Hruza G, Taub AF, Collier SL, Mulho SR (2009) Skin rejuvenation and wrinkle reduction using a fractional radiofrequency system. J Drugs Dermatol 8:259–265
3. Sadick N (2008) Tissue tightening technologies: fact or fiction. Aesthet Surg J 28(2):180–188
4. Mayoral FA (2007) Skin tightening with a combined unipolar and bipolar radiofrequency device. J Drugs Dermatol 6(2):212–215
5. Hantash BM, Bedi VP, Sudireddy V et al (2006) Laser-induced transepidermal elimination of dermal content by fractional photothermolysis. J Biomed Opt 11(4):041115
6. Habib N, Saedi N, Zachary C (2011) Cold-induced urticaria after fractional carbon dioxide laser resurfacing of the face. Dermatol Surg 37(11):1700–1703
7. Fife DJ, Zachary CB (2009) Delayed pinpoint purpura after fractionated carbon dioxide treatment in patient taking ibuprofen in the postoperative period. Dermatol Surg 35:553
8. Lowe NJ, Lask G, Griffin ME (1995) Laser skin resurfacing. Pre- and posttreatment guidelines. Dermatol Surg 21(12):1017–1019
9. Graber EM, Tanzi EL, Alster TS (2008) Side effects and complications of fractional laser photothermolysis: experience with 961 treatments. Dermatol Surg 34:301–305
10. Gotkin RH, Sarnoff DS, Cannarozzo G et al (2009) Ablative skin resurfacing with a novel microablative CO2 laser. J Drugs Dermatol 8:138–144
11. Setyadi HG, Jacobs AA, Markus RF (2008) Infectious complications after nonablative fractional resurfacing treatment. Dermatol Surg 34:1595–1598
12. Nanni CA, Alster TS (1998) Complications of carbon dioxide laser resurfacing. An evaluation of 500 patients. Dermatol Surg 24:315–320
13. Metelitsa AI, Alster TS (2010) Fractionated laser skin resurfacing treat-ment complications: a review. Dermatol Surg 36:299–306
14. Alam M, Pantanowitz L, Harton AM, Arndt KA et al (2003) A prospective trial of fungal colonization after laser resurfacing of the face: correlation between culture positivity and symptoms of pruritus. Dermatol Surg 29(3):255–260
15. Conn H, Nanda VS (2000) Prophylactic fluconazole promotes reepithelialization in full-face carbon dioxide laser skin resurfacing. Lasers Surg Med 26(2): 201–207
16. Rao J, Golden TA, Fitzpatrick RE (2002) Atypical mycobacterial infection following blepharoplasty and full-face skin resurfacing with CO2 laser. Dermatol Surg 288:768–771
17. Palm MD, Butterwick KJ, Goldman MP (2010) Mycobacterium chelonae infection after fractionated carbon dioxide facial resurfacing (presenting as an atypical acneiform eruption): case report and literature review. Dermatol Surg 36(9):1473–1481
18. Shamsaldeen O, Peterson JD, Goldman MP (2011) The adverse events of deep fractional CO2: a retrospective study of 490 treatments in 374 patients. Laser Surg Med 43(6):453–456
19. Ross RB, Spencer J (2008) Scarring and persistent erythema after fractionated ablative CO2 laser resurfacing. J Drugs Dermatol 7(11):1072–1073
20. Ramsdell WM (2012) Fractional CO2 laser resurfacing complications. Semin Plast Surg 26:137–140
21. Mamelak AJ, Goldberg LH, Marquez D, Hosler GA et al (2009) Eruptive keratoacanthomas on the legs after fractional photothermolysis: report of two cases. Dermatol Surg 35(3):513–518

Erratum to: Complications Associated with Botulinum Toxin Administration

Alexander Daoud and Martin Zaiac

Erratum to:

Chapter 13 in G. Fabbrocini et al. (eds.), *Nonsurgical Lip and Eye Rejuvenation Techniques*, DOI 10.1007/978-3-319-23270-6_13

Ivan Camacho is not a contributor to this chapter. Hence, the author's name and affiliation were removed. The updated authors names and affiliations should read as follows:

Alexander Daoud and Martin Zaiac

The online version of the updated chapter can be found at
DOI 10.1007/978-3-319-23270-6_13

A. Daoud, BS
Greater Miami Skin and Laser Center,
Florida International University, Miami, FL, USA

M. Zaiac, MD (✉)
Greater Miami Skin and Laser Center,
Florida International University,
Herbert Wertheim College of Medicine,
Miami Beach, FL, USA
e-mail: drmartyz@aol.com

G. Fabbrocini et al. (eds.), *Nonsurgical Lip and Eye Rejuvenation Techniques*,
DOI 10.1007/978-3-319-23270-6_15

Printed by Printforce, the Netherlands